Generation of Electrical Power

By
Dr. Hidaia Mahmood Alassouli

Introduction

This book includes my lecture notes for electrical power generation course. The layout, main components, and characteristics of common electrical power generation plants are described with application to various thermal power plants.

The book is divided to different learning outcomes

- CLO 1- Describe the layout of common electrical power generation plants.
- CLO 2- Describe the main components and characteristics of thermal power plants.

a) **CLO1 Describe the layout of common electrical power generation plants.**
- Explain the demand of base - power stations, intermediate - power stations, and peak- generation power stations.
- Describe the layout of thermal, hydropower, nuclear, solar and wind power generation plants.
- Identify the size, efficiency, availability and capital of generation for electrical power generation plants.
- Eexplain the main principle of operation of the transformer and the generator.

b) **CLO2: Describe the main components and characteristics of thermal power plants.**
- Identify the structure and the main components of thermal power plants.
- Describe various types of boilers and combustion process.
- List types of turbines, explain the efficiency of turbines, impulse turbines, reaction turbines, operation and maintenance, and speed regulation, and describe turbo generator.
- Explain the condenser cooling - water loop.
- Discuss thermal power plants and the impact on the environment.

Part 1: Describe the layout of common electrical power generation plants.

- Explain the demand of base - power stations, intermediate - power stations, and peak generation power stations.

- Describe the layout of thermal, hydropower, nuclear, solar and wind power generation plants.

- Identify the size, efficiency, availability and capital of generation for electrical power generation plants.

LO 1
Describe the layout of common electrical power generation plants

- Explain the demand of **base** - power stations, **intermediate** - power stations, and **peak**- generation power stations.
- Describe the layout of **thermal, hydropower, nuclear, solar** and **wind** power generation plants.
- Identify the **size, efficiency, availability** and **capital** of generation for electrical power generation plants.

1.1 Eexplaining the main principle of operation of the transformer and the generator.

Transformer

- The ideal transformer
- Vp ~=>ip ~ => Ø ~=> induced voltage in primary
- => induced voltage in secondary winding
- The induced voltage in the primary winding
- $v_p = N_p \frac{d\emptyset}{dt}$
- The induced voltage in the secondary winding
- $v_s = N_s \frac{d\emptyset}{dt}$
- $\frac{Vp}{Vs} = \frac{Np}{Ns} = a$

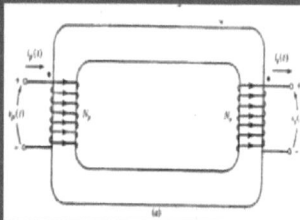

- The current equation
- $N_p i_p - N_s i_s = \emptyset R$
- $R = \frac{lc}{\mu A}$
- Reluctance of core very small =R=0
- $N_p i_p - N_s i_s = 0$
- $\frac{i_p}{i_s} = 1/a$

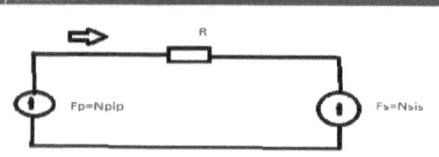

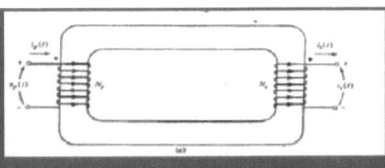

- The active power equation:

$$P_{in} = V_P I_P \cos \theta_P$$

$$P_{out} = V_S I_S \cos \theta_S$$

$$P_{out} = \frac{V_P}{a} a I_P \cos \theta$$

$$P_{out} = V_P I_P \cos \theta = P_{in}$$

- The reactive power equation

$$Q_{in} = V_P I_P \sin \theta = V_S I_S \sin \theta = Q_{out}$$

$$S_{in} = V_P I_P = V_S I_S = S_{out}$$

Referring the load to primary

- Referring the load from the secondary to primary

$$Z_L = \frac{V_S}{I_S}$$

$$V_P = aV_S$$

$$I_P = \frac{I_S}{a}$$

$$Z'_L = \frac{V_P}{I_P}$$

$$Z'_L = \frac{V_P}{I_P} = \frac{aV_S}{I_S/a} = a^2 \frac{V_S}{I_S}$$

$$Z'_L = a^2 Z_L$$

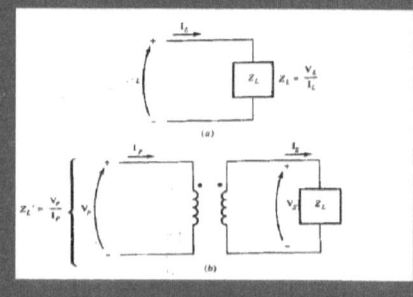

Three phase transformer

- Wye-Wye Connection

$$\frac{V_{LP}}{V_{LS}} = \frac{\sqrt{3}V_{\phi P}}{\sqrt{3}V_{\phi S}} = a$$

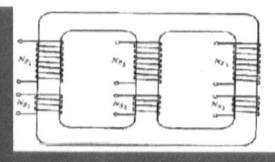

- Wye Delta Connection

$$\frac{V_{LP}}{V_{LS}} = \frac{\sqrt{3}V_{\phi P}}{V_{\phi S}}$$

$$\boxed{\frac{V_{LP}}{V_{LS}} = \sqrt{3}a \quad Y\text{-}\Delta}$$

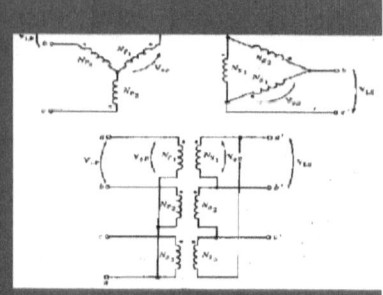

- Delta Wye Connection

$$\frac{V_{LP}}{V_{LS}} = \frac{V_{\phi P}}{\sqrt{3}V_{\phi S}}$$

$$\boxed{\frac{V_{LP}}{V_{LS}} = \frac{a}{\sqrt{3}} \quad \Delta\text{-}Y}$$

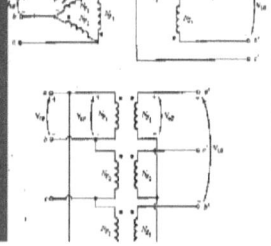

- Delta Delta Connection

$$V_{LP} = V_{\phi P}$$
$$V_{LS} = V_{\phi S}$$

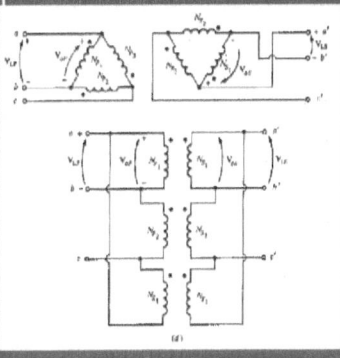

Synchronous Generator

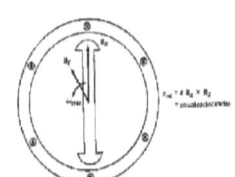

- Principle of Operation
- 1) The DC current applied to rotor winding produces magnetic field B_R
- 2) The rotor turned by primemover so there will be rotating magnetic field within the machine.
- 3) The rotating magnetic field induces 3 phase set of voltages within the stator windings.

Equivalent Circuit

- The relation between the speed of rotation and the frequency of the synchronous machine

$$f_e = \frac{n_m P}{120}$$

where f_e = electrical frequency, Hz
n_m = mechanical speed of magnetic field, r/min (= speed of rotor for synchronous machines)
P = number of poles

- The output voltage

$$E_A = K\phi\omega$$

- The equivalent circuit

$$V_\phi = E_A - jX_S I_A - R_A I_A$$

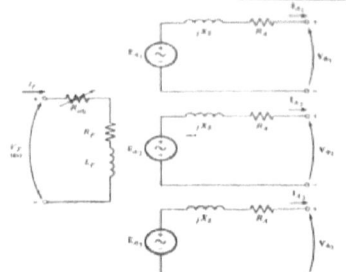

- The windings can be connected in Star or Delta

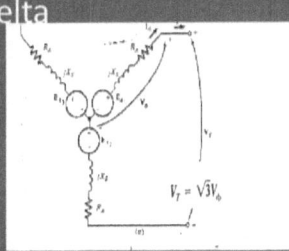
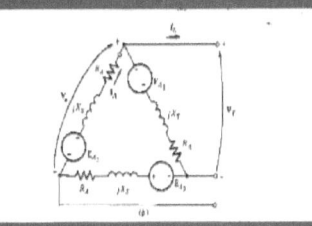

$V_T = \sqrt{3} V_\phi$

- The per phase equivalent circicuit

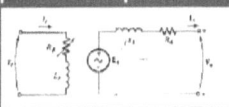

1.2. Explaining the demand of base - power stations, intermediate - power stations, and peak generation power stations.

SOURCES OF ELECTRICAL ENERGY

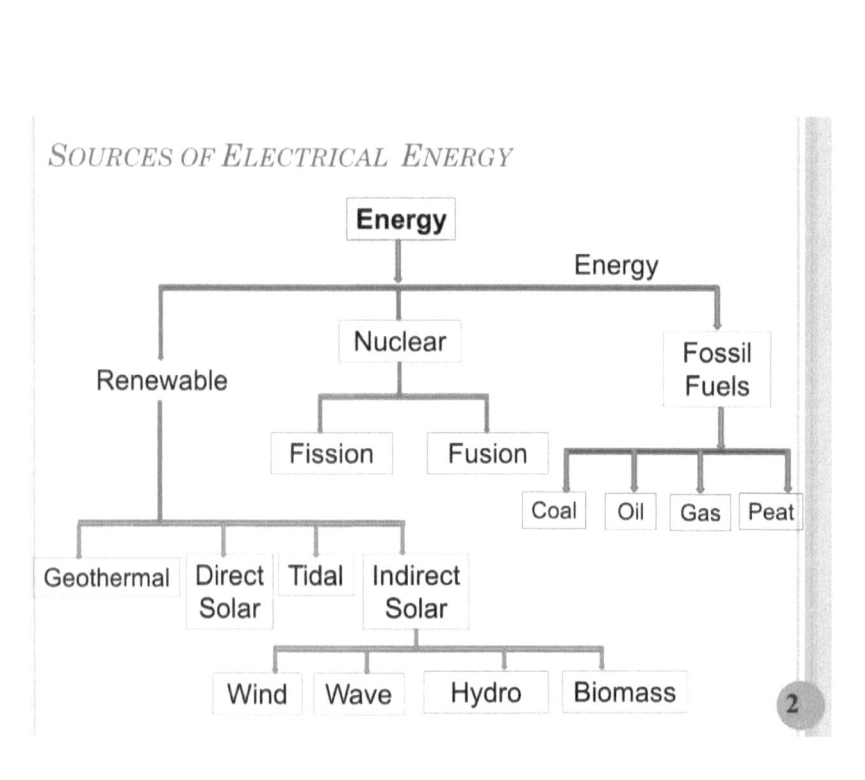

SOURCES OF ELECTRICAL ENERGY

The problems associated with the use of large quantities of energy are:

- ❖ Depletion of reserves
- ❖ Pollution and environmental degradation
- ❖ High financial cost
- ❖ Security of supply

WORLD ENERGY CONSUMPTION

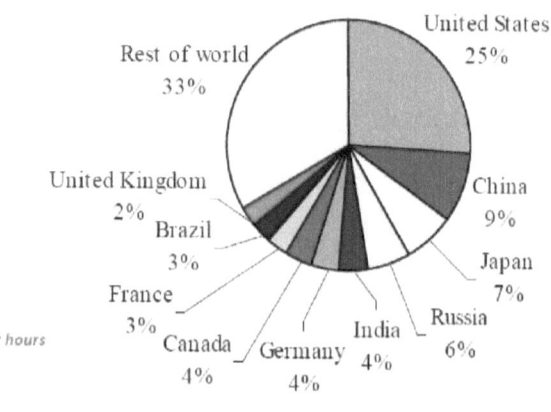

Consumed Electric Energy worldwide

- United States 25%
- Rest of world 33%
- China 9%
- Japan 7%
- Russia 6%
- India 4%
- Germany 4%
- Canada 4%
- France 3%
- Brazil 3%
- United Kingdom 2%

14000 Terawatt hours
$14.0 * 10^{12}$ kWh

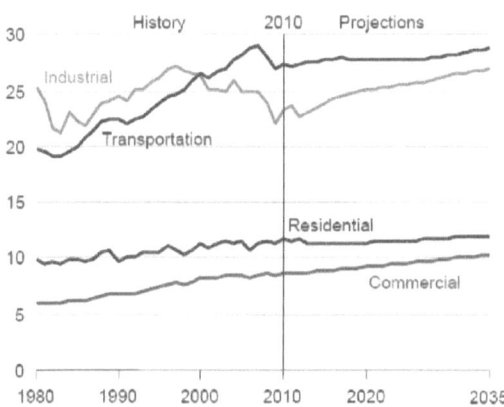
Delivered energy consumption by sector, 1980-2035 (quadrillion Btu)

Source: EIA, International Energy Outlook 2012

WORLD ENERGY CONSUMPTION

Energy production by fuel, 1980-2035 (quadrillion Btu)

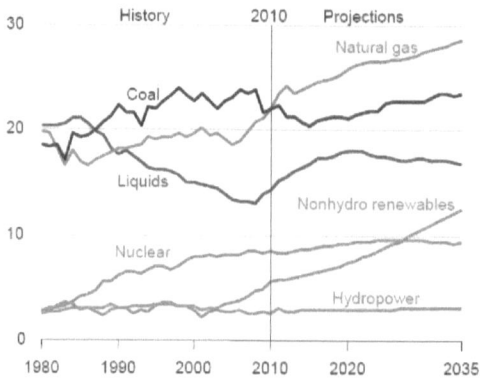

Source: EIA, International Energy Outlook 2012

Conservation of Energy

Using less energy can be accomplished in different ways:

1. **Technical Methods**
   Developing more efficient processes to achieve present conditions with less energy input.
2. **Adaptation of lifestyle and habits**
   Energy consumption could be reduced if individuals adopt different lifestyles
3. **Develop "Renewable" resources**

Introduction to Power Systems

Electricity is the main source of energy to keep life moving.

Some advantages of Electrical Energy:

1. Convenient: Easily converted to other forms: heat, mechanical, etc…
2. Easily controlled.
3. Flexible: Can be easily transmitted at long distances and within micro seconds (μs).
4. Relatively efficient.
5. Environmentally friendly.

BASIC COMPONENTS OF ELECTRICAL POWER SYSTEM

Modern power systems are made up of three *distinct* and *distant* components:

1 Generation

2 Transmission

3 Distribution

A Typical Electrical Power System

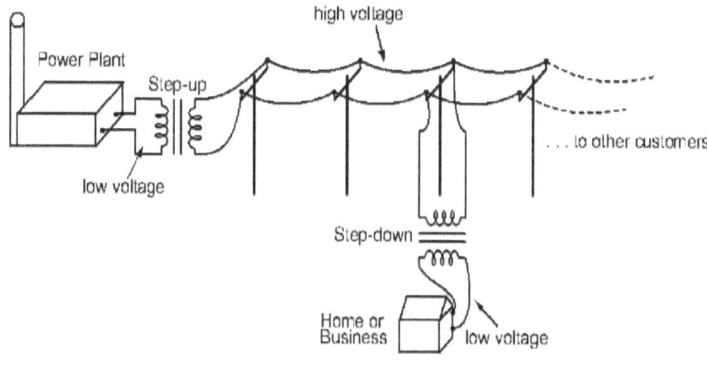

THE FUNCTION OF A POWER SYSTEM

Electrical energy cannot be stored; it must be generated when (but not necessarily where) it is needed.

The function of a power system is to meet the energy demand of the residential, commercial and industrial consumers connected to it, safely and reliably.

The Structure of a Power System

A power system, no matter how small, consists of the following parts:

- Generation plants

- Transmission system

- Substations to convert between different voltage levels

- Distribution system reaching all customers

REQUIREMENTS OF POWER SYSTEMS

A power system, no matter how simple, must meet the following requirements:

- The voltage and frequency at the consumer's premises is kept within declared limits
- The power losses in the system are small percentage
- The maximum current passing through the conductor is limited, so as not to cause overheating
- Comply with the safety standards and be environmental friendly

A Simple Power System

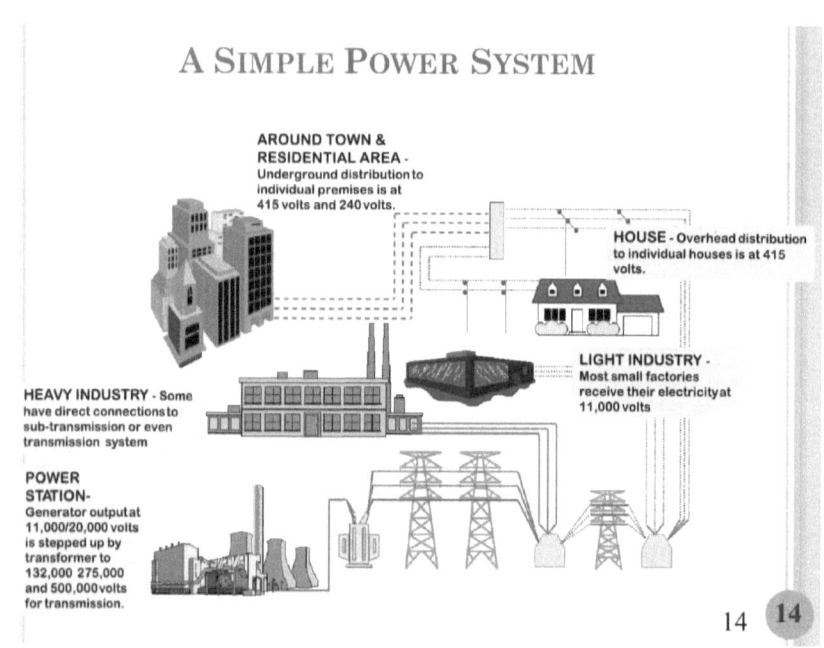

THE GRID SYSTEM

A grid is the combination of generating stations; the connecting transmission system; the conversion substations and distribution systems to deliver energy to various consumer centers.

Nowadays, different parts of power systems in countries or even continents are connected together to make a grid system

DISADVANTAGES OF THE GRID SYSTEM

- The increased generation in a grid system will result in large short circuit current. Circuit breakers interrupting capacities will be higher and therefore costlier.

- Failure of any part of the grid could result in a blackout in the whole grid.

A Grid System-Example

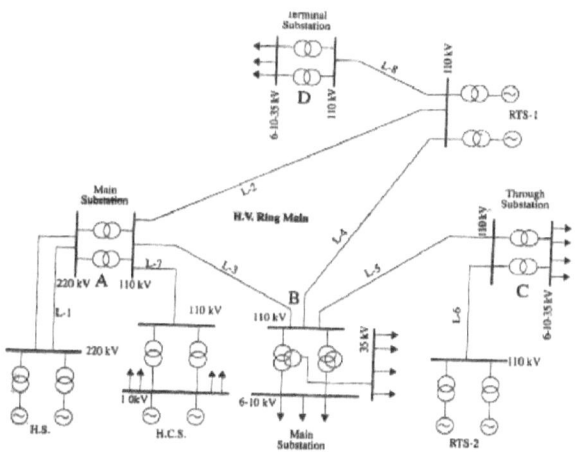

GENERATION

Basics of Electricity Generation:

Power Station is the name used for the plant where electricity is generated. It is the collection of all the equipment needed to generate electricity *safely, reliably and economically*. Examples of equipment that can be found in a power station are: *generators, turbines, water and fuel pumps, boilers, diesel engines* etc.

DAILY VARIATION OF POWER DEMAND CURVES

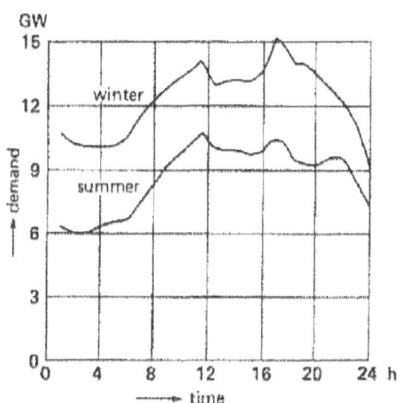

***A power demand curve of typical winter & summer
days in North America***

ELECTRICAL ENERGY GENERATION BY ADWEA

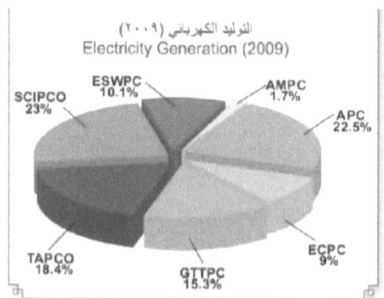

Electricity Generation (2009) — التوليد الكهربائي (٢٠٠٩)
- ESWPC 10.1%
- SCIPCO 23%
- AMPC 1.7%
- APC 22.5%
- ECPC 9%
- GTTPC 15.3%
- TAPCO 18.4%

Global Electricity Capacity & Generation (2009) — إجمالي القدرة الانتاجية و الطاقه المولده (٢٠٠٩)

Company	Capacity (MW)	Generation (GWh)	الشركة
AMPC	550	759	المرفأ
APC	2,433	9,795	العربية
ECPC	759	3,899	الإمارات
GTTPC	1,672	6,661	الخليج
SCIPCO	1,615	10,041	الشويهات
TAPCO	2,220	8,034	تابكو
ESWPC	861	4,426	سيمكورب
Total *	10,110	43,615	الإجمالي*
Takreer & EMAL's Net Imports	-	183	تكرير وإيمال

* Excluding Takreer

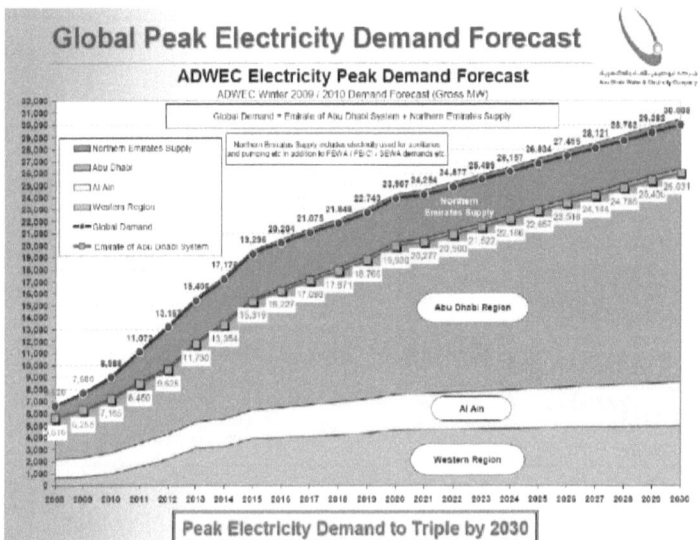

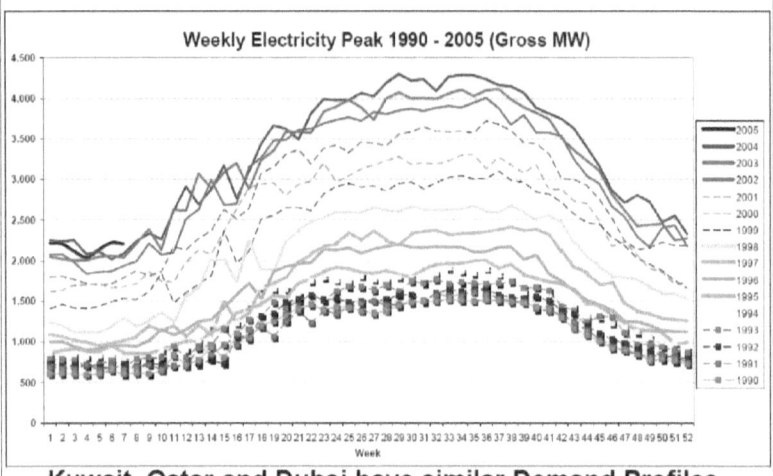

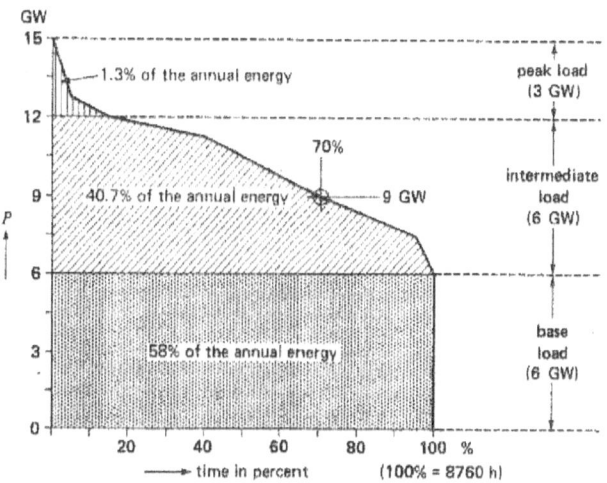

BASE LOAD, INTERMEDIATE LOAD AND PEAK LOAD

Base Load:

This is the load demanded 100% of the time

Peak Load:

This is the load that is demanded for a brief intervals during the day.

Intermediate Load:

The area between the peak load and the base load is the intermediate load

PRESENTATION ASSIGNMENT (2)

- Gas turbine power plant
- Combined cycle power plant
- Solar power plant
- Geothermal power plant

- Deadline: 13-03-2015

PRESENTATION ASSIGNMENT (2)

- Principle of operation
- Main parts and their functions
- Power plant layout
- Schematic diagram
- Advantages and disadvantages
- Environmental impacts

POWER STATION TYPES

Station Type	Description	Examples
Base	To deliver full power at all times	Thermal power stations Nuclear and coal, oil, or gas–fired power stations
Intermediate	To respond quickly to changes in demand by adding or removing one or more generating units	Hydro power plants and Gas turbines
Peak	To deliver power for short intervals at a time. Such power stations must be put into service very quickly	Diesel engines, gas turbines, pumped-storage

1.3. Describing the layout of thermal, hydropower and wind power generation plants.

TYPES OF POWER STATIONS

There are FOUR basic types of generating stations:

- **Thermal generating stations**
- **Hydropower generating stations**
- **Nuclear generating stations**
- **Renewable Energy**

1- Thermal (Steam) Power Stations

Thermal power stations convert water into steam at high pressure.

Steam is then directed at high speed at the blades of steam turbines causing it to rotate.

The generator, which is coupled to the turbine, will rotate and generate electrical energy.

Thermal Power Plant

1. First the pulverized coal is burnt into the <u>furnace of steam boiler</u>.
2. High pressure steam is produced in the boiler.
3. This steam is then passed through the super heater, where it further heated up.
4. This supper heated steam is then entered into a turbine at high speed.
5. In turbine this steam force rotates the turbine blades that means here in the turbine the stored potential energy of the high pressured steam is converted into mechanical energy.
6. After rotating the turbine blades, the steam has lost its high pressure, passes out of turbine blades and enters into a condenser.
7. In the condenser the cold water is circulated with help of pump which condenses the low pressure wet steam.
8. This condensed water is then further supplied to low pressure water heater where the low pressure steam increases the temperature of this feed water, it is then again heated in a high pressure heater where the high pressure of steam is used for heating.
9. The turbine in thermal power station acts as a prime mover of the alternator.

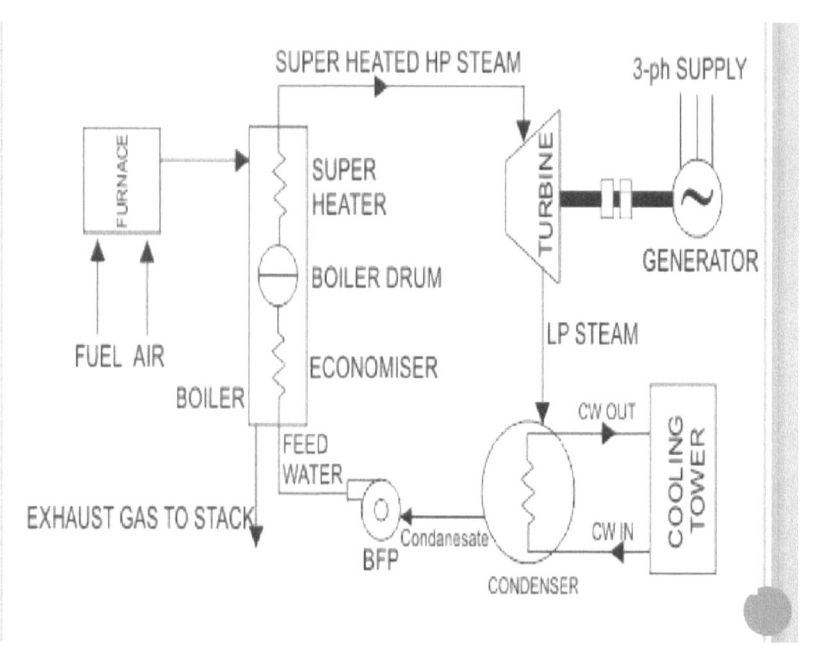

THERMAL POWER STATION

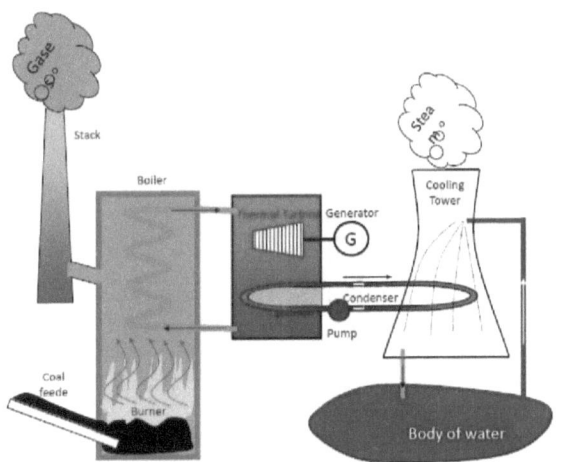

THERMAL POWER STATION

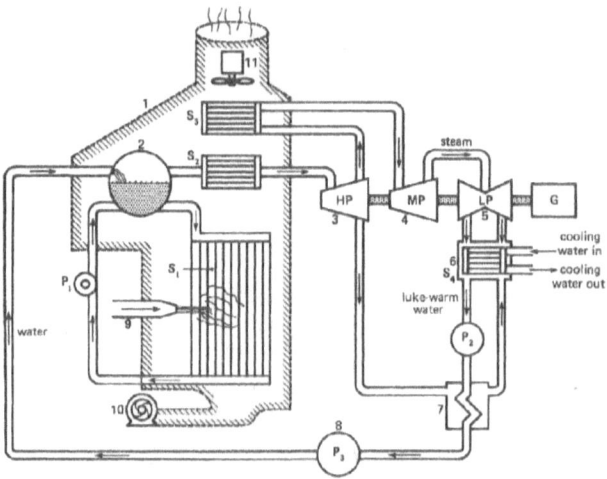

Schematic diagram of a thermal power station

THERMAL POWER STATION
Major Parts

1 Boiler:
This is a large structure containing the burner and the furnace. The furnace is surrounded by large number of tubes (in water tube boilers) in which water is circulated at all times by pump P1.

2 Drum (Steam/Water Tank):
This is the vessel in which the steam, generated in the boiler tubes, is collected. The lower part of this vessel contains the supply water for the boiler tubes. The steam in this vessel is saturated and is not suitable to be injected onto the turbine blades.

S1- The boiler vessel in which water is circulated (in water tubes boiler) and around which the fire is distributed so as to cause maximum water heating.

THERMAL POWER STATION
Major Parts

S2 & S3- Superheaters:
The steam produced in the boiler is saturated (wet), which makes it inefficient and can damage the turbine blades. So, after being produced in the boiler tank (2), the steam is passed to superheaters (S2 & S3), to produce superheated and dry steam.

3, 4 & 5 – Steam Turbine High Pressure (HP), Medium Pressure (MP) and Low Pressure (LP) stages:
The steam coming out of the boiler and superheater S1 is ready to go into stage 1 (HP) of the turbine. The steam coming out of stage 1 of the turbine is superheated before it can be injected into stage 2 (MP) of the turbine.

THERMAL POWER STATION
Major Parts

S4 – Condenser:
The steam coming out of stage 3 (LP) of the turbine would have lost most of it's heat and pressure and is ready to be taken back into the boiler after being condensed in the condenser.

7- Preheater (Economiser):
To increase boiler efficiency, supply water is usually directed to preheater (7) which receives hot steam from the boiler and use it to preheat water before it is fed back into the boiler.

Steam Turbine:
Dry and superheated steam from the boiler and superheater is directed to the various stages of the turbine [3 (HP), 4 (MP) & 5 (LP)].

2- Gas Turbine Power Plant

1. Here the air is first compressed at desired pressure then it is brought to a combustion chamber where the compressed air is heated up by means of fuel combustion.

2. Then this highly compressed hot air is released from the combustion chamber through the nozzle to a turbine, called gas turbine.

3. During expansion of pressurized and hot air, mechanical work is done to rotate the turbine. As the turbine rotates, the alternator also rotates since a common shaft is shared by both turbine and alternator in the gas turbine power plant.

4. It is needed to be mentioned here that not only the turbine and alternator, the air compressor is also fitted on the same shaft.

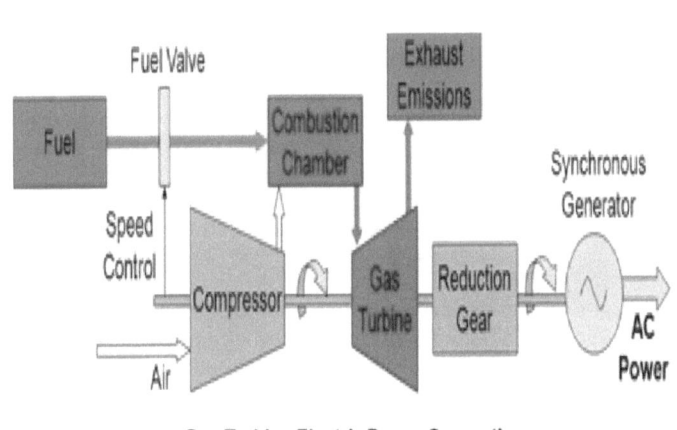

Gas Turbine Electric Power Generation

3- Combined Cycle Power Plant

Various steps in combined cycle power plant are the following:

Atmospheric air (A) is compressed in compressor and pressurized air (B) enters into the GT combustor where it mixes with fuel and undergoes combustion.

The resulting very high temperature gases (C) enter into gas turbine.

The exhaust gases (D) from gas turbine have temperature of around 500 - 550°C. Heat recovery steam generator (HRSG) generates steam using water (E) by utilizing exhaust gases and finally, flue gases from HRSG are sent to stack.

The generated steam (F) drives steam turbine and exhaust from steam turbine is fed to the condenser. Condensed water is again pumped back to the HRSG.

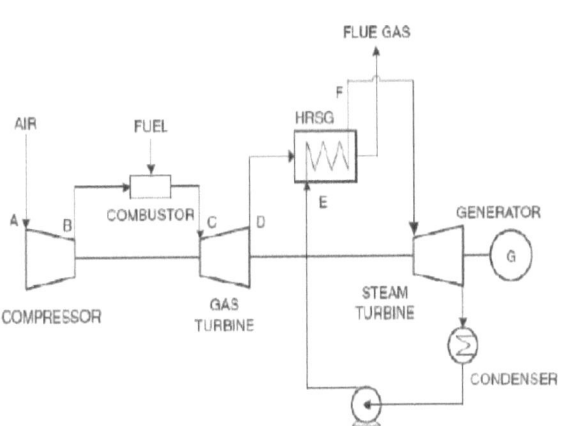

4-Hydroelectric Power Stations

Hydroelectric power plants use no fuel! They use the kinetic energy of falling water. Water is collected behind high dams, and when released it strikes the blades of the turbine, causing it to rotate. This will cause the generator, which is coupled to the turbine, to rotate and generate electricity. As coal, oil and uranium are becoming more expensive, hydro plants are becoming more popular.

HYDROELECTRIC POWER STATIONS

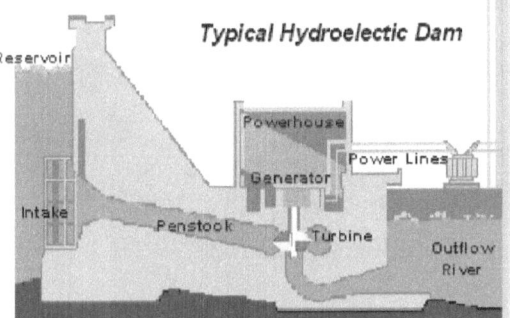

http://static.howstuffworks.com/gif/hydropower-plant-usbr-hoover.jpg

http://ga.water.usgs.gov/edu/hyhowworks.html

Construction and Working of Hydro Power Plant

Fundamental parts of hydro power plant are
- **Dam**
- **Reservoir**
- **Penstock**
- **Storage tank**
- **Turbines and generators**
- **Switchgear and protection**

Hydroelectric power plant

The typical layout of a hydroelectric power plant and its basic components.

- Dam and Reservoir: The dam is constructed on a large river in hilly areas to ensure sufficient water storage at height. The dam forms a large reservoir behind it. The height of water level (called as water head) in the reservoir determines how much of potential energy is stored in it.

- Control Gate: Water from the reservoir is allowed to flow through the penstock to the turbine. The amount of water which is to be released in the penstock can be controlled by a control gate. When the control gate is fully opened, maximum amount of water is released through the penstock.

- Penstock: A penstock is a huge steel pipe which carries water from the reservoir to the turbine. Potential energy of the water is converted into kinetic energy as it flows down through the penstock due to gravity.

- Water Turbine: Water from the penstock is taken into the water turbine. The turbine is mechanically coupled to an electric generator. Kinetic energy of the water drives the turbine and consequently the generator gets driven. There are two main types of water turbine; (i) Impulse turbine and (ii) Reaction turbine. Impulse turbines are used for large heads and reaction turbines are used for low and medium heads.

- <u>Generator</u>: A generator is mounted in the power house and it is mechanically coupled to the turbine shaft. When the turbine blades are rotated, it drives the generator and electricity is generated which is then stepped up with the help of a <u>transformer</u> for the transmission purpose.

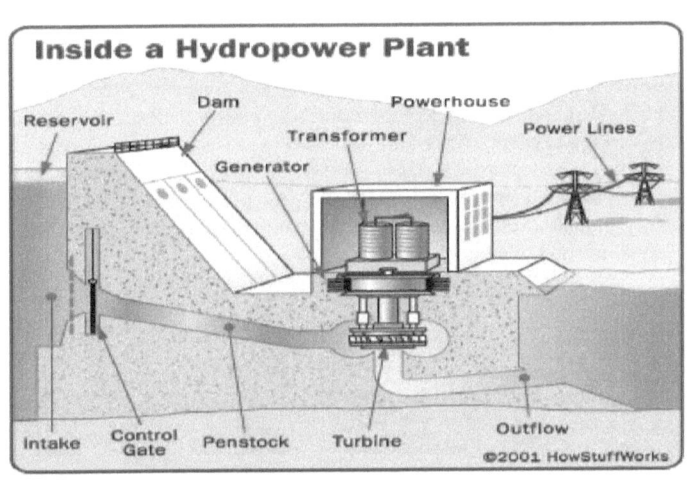

- **Surge Tank:** Surge tanks are usually provided in high or medium head power plants when considerably long penstock is required. A surge tank is a small reservoir or tank which is open at the top. It is fitted between the reservoir and the power house. The water level in the surge tank rises or falls to reduce the pressure swings in the penstock. When there is sudden reduction in load on the turbine, the governor closes the gates of the turbine to reduce the water flow. This causes pressure to increase abnormally in the penstock. This is prevented by using a surge tank, in which the water level rises to reduce the pressure. On the other hand, the surge tank provides excess water needed when the gates are suddenly opened to meet the increased load

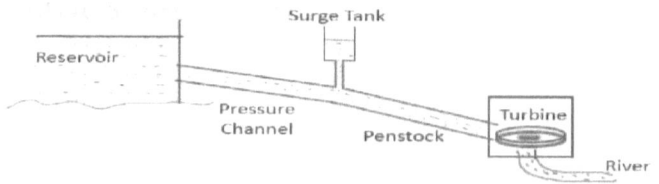

HYDROELECTRIC POWER STATIONS

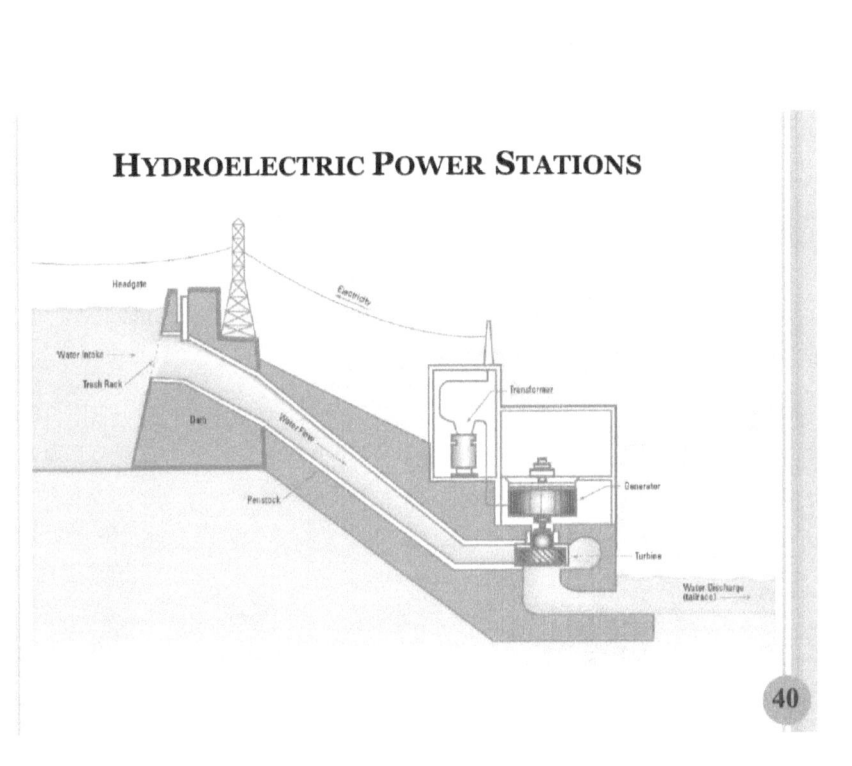

CLASSIFICATION OF HYDROPOWER PLANTS

The hydropower plants are classified according to the head of water.
- If the operating head of water exceeds 70 meters, the plant is known as "high head power plant". Pelton turbine is used as prime mover in this type of power plants.

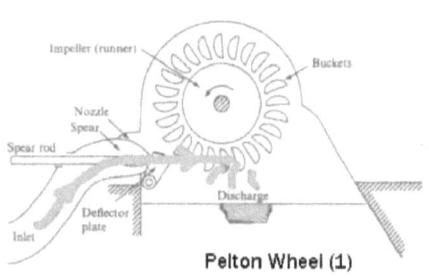

Pelton Wheel (1)

CLASSIFICATION OF HYDROPOWER PLANTS

- If the operating head of water ranges from 15 to 70 meters then the power plant is known as medium head power plant.
- If the operating head of water is less than 15 meters that power plant is known as low head power plant.

CLASSIFICATION OF HYDROPOWER PLANTS

Large-hydro	More than 100 MW and usually feeding into a large electricity grid
Medium hydro	15 - 100 MW - usually feeding a grid
Small-hydro	1 - 15 MW - usually feeding into a grid
Mini-hydro	Above 100 kW, but below 1 MW; either stand alone schemes or more often feeding into the grid
Micro-hydro	From 5kW up to 100 kW; usually provided power for a small community or rural industry in remote areas away from the grid.

CLASSIFICATION OF HYDROPOWER PLANTS

Based on the way energy is extracted, there are many types of hydropower including the following:

- Impoundment hydropower plants
- Diversion / run-off-river hydropower plants
- Pumped storage hydropower plants
- Micro hydropower plants

Impoundment hydropower plants

- The most common type of hydroelectric power plant is an impoundment facility.
- An impoundment facility, typically a large hydropower system, uses a dam to store river water in a reservoir.
- Water released from the reservoir flows through a turbine, spinning it, which in turn activates a generator to produce electricity.
- The water may be released either to meet changing electricity needs or to maintain a constant reservoir level.

Impoundment hydropower plants

Impoundment hydropower plants

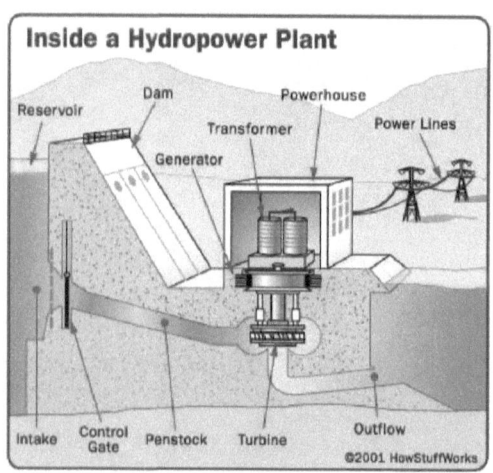

River run off hydropower plants

A diversion, sometimes called run-of-river, facility channels a portion of a river through a canal or penstock. It may not require the use of a dam.

River run off hydropower plants

These plants use the natural downward flow of rivers and to capture the kinetic energy carried by water. Typically water is taken from the river at a high point and gravity fed down a pipe to a lower point where it emerges through a turbine generator and re-enters the river. Installation of such a system is relatively cheap and has very little environmental impact.

River run off hydropower plants

Pumped-storage hydropower plants

Pumped-Storage Hydropower Plants

Pumped-storage power plants consists of two lakes, one at the foot of a high mountain and the other is on top of the mountain. Tunnels are dug in the mountain between the two lakes. A set of generators/motors are built at the foot of the mountain. During low power demand (after midnight), water is pumped from the lower lake to the top lake. When the peak hour approaches (in the early morning), the pumps are stopped and water is allowed to drop from the top lake to the bottom lake, causing the generator to produce electrical energy to supply peak hour demand.

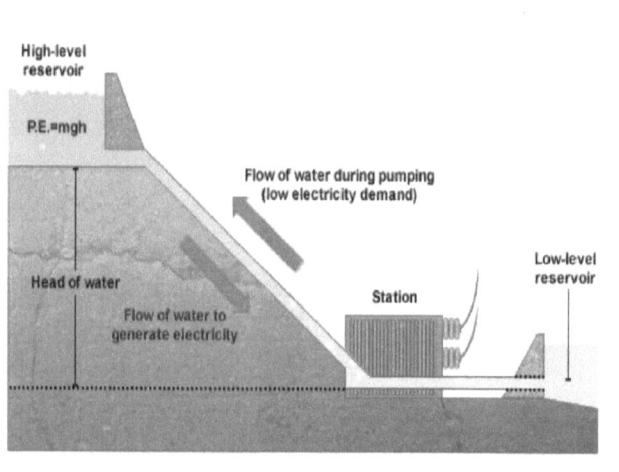

Micro hydropower plants

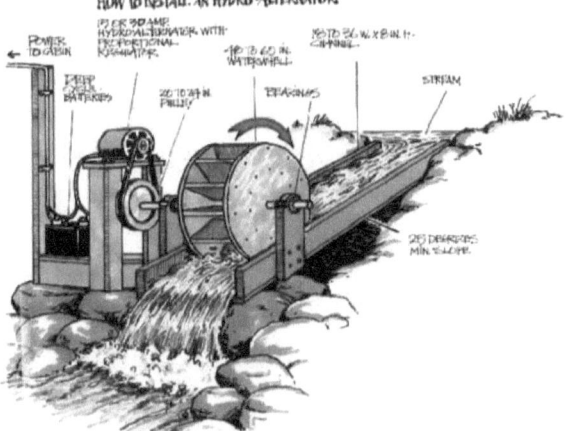

Advantages of Hydropower Stations

- Water is used as the source of fuel, which cheaper and easily available
- No smoke, no ashes or atmospheric pollution.
- Running charges are small.
- Availability time is very small which makes hydro electric power plants suitable for peak power station duties.
- In addition to electrical energy generation, such stations help in flood control and irrigation.

Disadvantages of Hydro Electric Power Stations

- High initial cost, mainly used for dam construction
- Climate dependent
- Fish killing
- High cost of transmission since hydro electric power plants are built well away from load centers.

Hydro Electric Power Stations

http://www.youtube.com/watch?v=cEL7yc8R42k

http://www.youtube.com/watch?v=wvxUZF4lvGw

WIND TURBINE

Advantages of Hydropower Stations

- Water is used as the source of fuel, which cheaper and easily available
- No smoke, no ashes or atmospheric pollution.
- Running charges are small.
- Availability time is very small which makes hydro electric power plants suitable for peak power station duties.
- In addition to electrical energy generation, such stations help in flood control and irrigation.

Disadvantages of Hydro Electric Power Stations

- High initial cost, mainly used for dam construction
- Climate dependent
- Fish killing
- High cost of transmission since hydro electric power plants are built well away from load centers.

HYDRO ELECTRIC POWER STATIONS

http://www.youtube.com/watch?v=cEL7yc8R42k

http://www.youtube.com/watch?v=wvxUZF4lvGw

WIND TURBINE

Wind Turbine

When the wind strikes the rotor blades, blades start to rotating. Rotor is directly connected to high speed gearbox. Gearbox converts the rotor rotation into high speed which rotates the electrical generator. An exciter is needed to give the required excitation to the coil so that it can generate required voltage. The exciter current is controlled by a turbine controller which senses the wind speed based on that it calculate the power what we can achieve at that particular wind speed.

Then output voltage of electrical generator is given to a rectifier and rectifier output is given to line converter unit to stabilize the output AC that is feed to the grid by a high voltage transformer.

Nacelle of Wind Turbine

Nacelle is a big box or kiosk that sits on the tower and houses all the components of a wind turbine. It houses electrical generator, power converter, gearbox, turbine controller, cables, yaw drive.

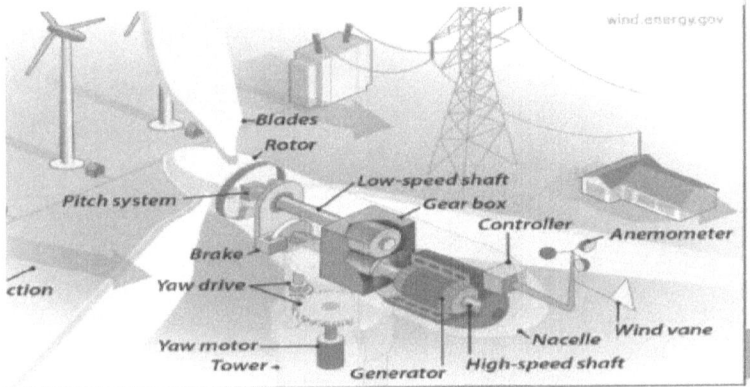

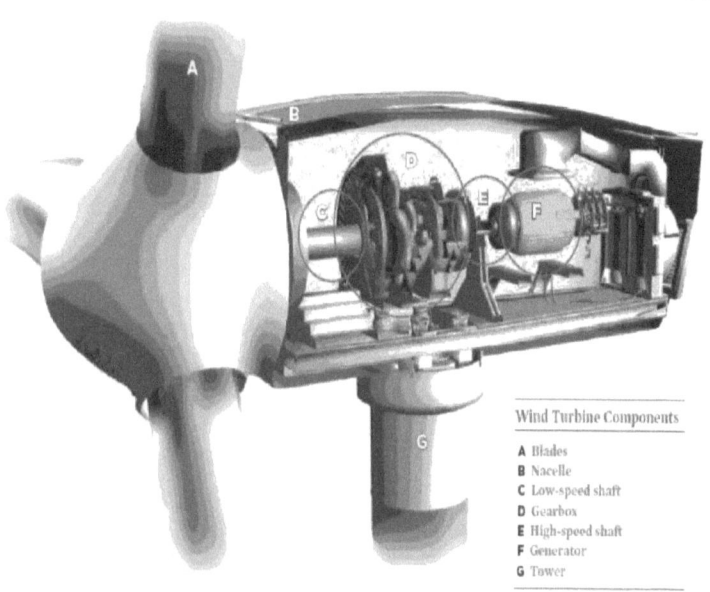

Wind Turbine Components

- **A** Blades
- **B** Nacelle
- **C** Low-speed shaft
- **D** Gearbox
- **E** High-speed shaft
- **F** Generator
- **G** Tower

Rotor Blades of Wind Turbine
Blades are the main mechanical parts of wind turbine. The blades convert wind energy into usable mechanical energy. When the wind strikes on the blades, the blades rotate. This rotation transfers its mechanical energy to the shaft. Blades are shaped like airplane wings.

Shaft of Wind Turbine
The shaft directly connected to the hub, is low speed shaft. When the blades rotate, this shaft spins with same rpm as the rotating hub. In most cases, the low speed main shaft is geared with a high speed shaft through a gearbox. In this way, the rotor blades transfers its mechanical energy to the shaft which

Gearbox
The wind turbine does not rotate in high speed rather it rotates gently in low speed. Gearbox increases the speed to much higher value. For example, if gearbox ratio is 1:80 and if the rpm of low speed main shaft is 15, the gearbox will increase the speed of generator shaft to $15 \times 80 = 1200$ rpm.

Power Converter
Because wind is not always constant so electrical potential generated from generator is not constant but we need a very stable voltage to feed the grid. Power converter is an electrical device that stabilizes the output alternating voltage transferred to the grid.

Turbine Controller
Turbine controller is a computer (PLC) that controls the entire turbine. It starts and stops the turbine and runs self diagnostic in case of any error in the turbine.

Anemometer
It measures the wind speed and passes the speed information to PLC to control the turbine power.

Wind Vane
It senses the direction of wind and passes the direction to PLC then PLC faces the blades in such a way that it cuts the maximum wind.

Pitch Drive
Pitch drive motors control the angle of blades whenever wind changes it rotates the angle of blades to cut the maximum wind, which is called pitching of blades.

Yaw Drive
Blades and other components in wind turbine is housed in nacelle, whenever any change in wind direction is there, the nacelle has to face in the direction of wind to extract the maximum energy from wind. For this purpose yaw drive motor are used to rotate the nacelle. It is controlled by PLC that uses the wind vane information to sense the wind direction.

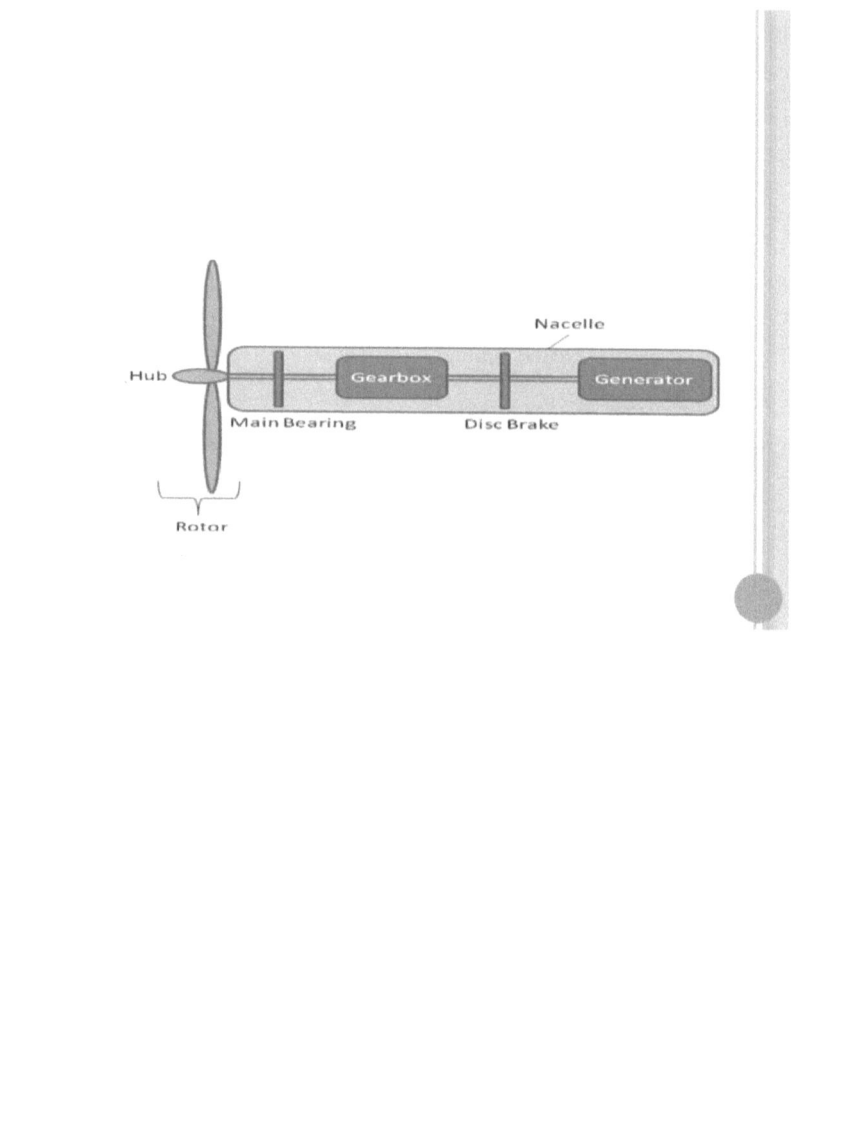

Wind Power

- Wind Power depends on:
  - amount of air (volume)
  - speed of air (velocity)
  - mass of air (density)

  flowing through the area of interest (flux)

- **Kinetic Energy** definition:
  - $KE = \frac{1}{2} * m * v^2$

- Power is KE per unit time:
  - $P = \frac{1}{2} * \dot{m} * v^2$

  $\dot{m} = \dfrac{dm}{dt}$ mass flux

- Fluid mechanics gives **mass flow rate** (density * volume flux):
  - $dm/dt = \rho * A * v$

- Thus:
  - $P = \frac{1}{2} * \rho * A * v^3$

  ⇒
  - Power ~ cube of velocity
  - Power ~ air density
  - Power ~ rotor swept area $A = \pi r^2$

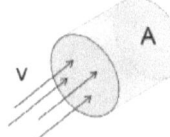

From, the above equation it is found the theoretically maximum power extracted from the wind is in the fraction of 0.5925 of it's total kinetic power. This fraction is known as Betz Coefficient. This calculated power is according to **theory of wind turbine** but actual mechanical power received by the generator is lesser than that and it is due to losses for friction rotor bearing and inefficiencies of aerodynamic design of the turbine.

From equation (4) it is clear that the extracted power is

Directly proportional to air density ρ. As air density increases, the power of the turbine increases.

Directly proportional to swept area of the turbine blades. If the length of the blade is increased, the radius of the swept area increases accordingly so turbine power increases.

Betz Limit & Power Coefficient:

- Power Coefficient, **Cp**, is the ratio of power extracted by the turbine to the total contained in the wind resource $C_p = P_T/P_W$
- Turbine power output

    $P_T = \frac{1}{2} * \rho * A * v^3 * C_p$

- The **Betz Limit** is the maximal possible $C_p = 16/27$
- 59% efficiency is the **BEST** a conventional wind turbine can do in extracting power from the wind

Wind Energy – The Facts

- Local wind speed is also an important factor since:

   power α (wind speed)^3

   power α (blade diameter)^2

- The local wind speed should be, on average, at least 7 m/s at 25 m above the earth's surface in order to make harnessing wind from it worthwhile.

Example: A wind turbine has a blade circle diameter of 35 m and is operating with a wind speed of 15 m/s. If the energy conversion efficiency is 90% and the density of air is 1.2 kg/m³, calculate the average power output of the turbine.

$$P = \eta \frac{1}{2} \rho A v^3$$
$$= \eta \frac{1}{2} \rho v^3 \pi \left(\frac{d}{2}\right)^2$$
$$= \frac{1}{8} \eta \rho v^3 \pi d^2$$

$P = 0.125 \times 0.9 \times 1.2 \times 15^3 \times 3.14 \times 35^2$
$= 1752561.5 \ W$
$\quad\quad = 1.75 \times 10^6 \ W = 1.75 \ MW$

1.4. Describing the layout of Nuclear Power Station and the Solar Power Plants.

ماذا نعرف على الانشطار النووي؟

الشكل التالي يوضح فكرة الانشطار النووي لنواة ذرة يورانيوم-235 حيث ان النواة تمتص النيوترون الساقط على النواة، وبمجرد امتصاص النيوترون تنقسم ذرة اليورانيوم-235 إلى ذرتين وتنطلق نيوترونين أو ثلاث نيوترونات جديدة وتنطلق طاقة من الذرتين الناتجتين عن الانشطار في صورة اشعة جاما وتعمل النيوترونات المحررة في الانشطار على الاصطدام بأنوية يورانيوم-235 أخرى وبتكرر هذه العملية في انشطار نووي متسلسل.

تحدث عملية امتصاص النيوترونات والانشطار النووي لليورانيوم-235 بسرعة كبيرة جداً حيث لا تستغرق هذه العملية اكثر من بيكوثانية، أي (1×10^{-12}) ثانية. وخلال فترة زمنية صغيرة جداً تحصل على طاقة هائلة تنطلق في صورة حرارة واشعاعات جاما وتطلق تسأل عزيزي القارئ من أين أتت هذه الطاقة الهائلة؟ ان الاجابة هي عبدأ بسيط لكن قانون بكافئ الطاقة والكتلة لاينشتين وهو ان الطاقة تساوي حاصل ضرب الكتلة في مربع سرعة الضوء وبالتالي أي كتلة صغيرة تضربها في مربع سرعة الضوء يؤدي الى طاقة هائلة ويكتب قانون بكافئ الطاقة والكتلة

$$E = mc^2$$

 تحول الى طاقة في الانشطار النووي لليورانيوم-235 يأتي من ان كتلة النواة الأم اكبر من كتلة نواة الذرتين المنشطرتين وبالتالي فرق الكتلة هذا هو مصدر الطاقة الهائلة التي تتولد عن الانشطار النووي لليورانيوم-235 واليس بقدر بحوالي 200 مليون الكترون فولت من طاقة تنجز من كل ذرة يورانيوم-235 ولنتخيل كم الذرات التي تكون في قطعة من اليورانيوم بحجم كرة تنس ولنتصور كم الطاقة الهائلة المتحرر من انشطارات في ذرات اليورانيوم في هذا الحجم الصغير فهذه يعادل انفجار 20 مليون لتر من الوقود.

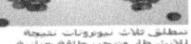

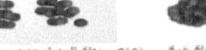

Different Components of Nuclear Power Station

A nuclear power station has mainly four components.

- Nuclear reactor,
- Heat exchanger,
- Steam turbine,
- Alternator.

Nuclear Reactor

In nuclear reactor, Uranium 235 is subjected to nuclear fission. It controls the chain reaction that starts when the fission is done. The chain reaction must be controlled otherwise rate of energy release will be fast, there may be a high chance of explosion. In nuclear fission, the nuclei of nuclear fuel, such as U^{235} are bombarded by slow flow of neutrons. Due to this bombarding, the nuclei of Uranium is broken, which causes release of huge heat energy and during breaking of nuclei, number of neutrons are also emitted. **This release of energy is due to mass defect. That means, the total mass of initial product would be reduced during fission. This loss of mass during fission is converted into heat energy as per famous equation $E = mc^2$, established by Albert Einstein.**

These emitted neutrons are called fission neutrons. These fission neutrons cause further fission. Further fission creates more fission neutrons which again accelerate the speed of fission. This is cumulative process. If the process is not controlled, in very short time the rate of fission becomes so high, it will release so huge amount of energy, there may be dangerous explosion. This cumulative reaction is called chain reaction. This chain reaction can only be controlled by removing fission neutrons from nuclear reactor. The speed of the fission can be controlled by changing the rate of removing fission neutrons from reactors.

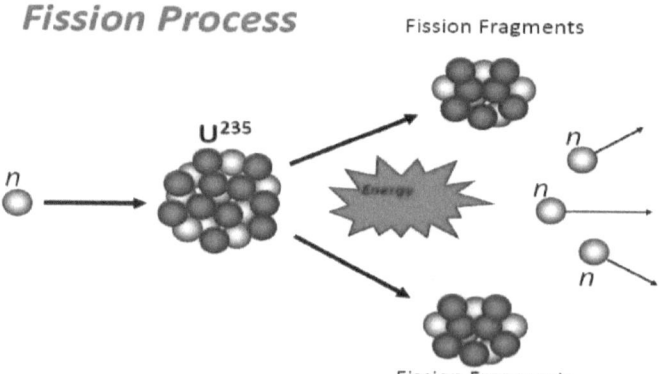

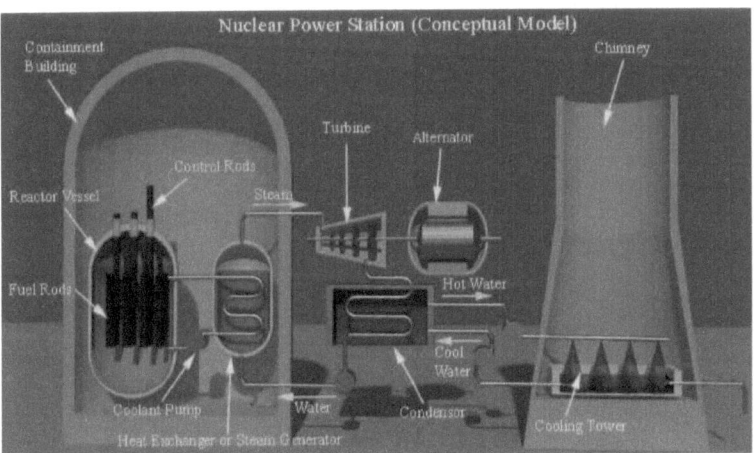

- A <u>nuclear reactor:</u> is a cylindrical shaped stunt pressure vessel. The fuel rods are made of nuclear fuel i.e. Uranium moderates, which is generally made of graphite cover the fuel rods. The moderates slow down the neutrons before collision with uranium nuclei. The controls rods are made of cadmium because cadmium is a strong absorber of neutrons.

The control rods are inserted in the fission chamber. These cadmium controls rods can be pushed down and pull up as per requirement. When these rods are pushed down enough, most of the fission neutrons are absorbed by these rods, hence the chain reaction stops. Again, while the controls rods are pulled up, the availability of fission neutrons becomes more which increases the rates of chain reaction. Hence, it is clear that by adjusting the position of the control rods, the rate of nuclear reaction can be controlled and consequently the <u>generation of electrical power</u> can be controlled as per load demand. In actual practice, the pushing and pulling of control rods are controlled by automatic feedback system as per requirement of the load. It is not controlled manually. The heat released during nuclear reaction, are carried to the heat exchanger by means of coolant consist of sodium metal.

- **Heat Exchanger:** In heat exchanger, the heat carried by sodium metal, is dissipated in water and water is converted to high pressure steam here. After releasing heat in water the sodium metal coolant comes back to the reactor by means of coolant circulating pump.

- **Steam Turbine:** In nuclear power plant, the steam turbine plays the same role as coal power plant. The steam drives the turbine in same way. After doing its job, the exhaust steam comes into steam condenser where it is condensed to provide space to the steam behind it.

- **Alternator:** An alternator, coupled with turbine, rotates and generates electrical power, for utilization. The output from alternator is delivered to the bus-bars through transformer, circuit breakers and isolators.

Control Rods:

Control rods play a vital role in the control of the energy generated in a nuclear reactor. Their role is to absorb the access neutrons in the reactor, and therefore slows down or even stops the fission process. This needs to be done when no electrical energy is needed or in case of emergency. To maintain a sustained controlled nuclear reaction, for every 2 or 3 neutrons released, only one must be allowed to strike another uranium nucleus.

If this ratio is less than 1, the reaction will die out; if it is greater than one it will grow uncontrolled (an atomic explosion). A neutron absorbing element must therefore be present to control the amount of free neutrons in the reaction space. Such element is called control rod. Most reactors are controlled by means of control rods that are made of a strongly neutron-absorbent material such as boron or cadmium.

Properties of the Control Rods:
- One property which is a must for control rod material is the heavy absorption capacity for neutrons so that they can carry out the control function effectively.
- Another property of control rods is that the rod material should not start a fission reaction despite the heavy absorption of neutrons.

- The Moderator:

In addition to the need to *capture* neutrons, the neutrons often have too much kinetic energy. These *fast neutrons* are slowed down through the use of a moderator such as heavy water and ordinary water. Some reactors use graphite as a moderator, but this design has problems. Once the fast neutrons have been slowed down, they are more likely to produce further nuclear fissions or be absorbed by the control rod.

- LIGHT WATER REACTORS(LWR)

The family of nuclear reactors known as light water reactors (LWR), cooled and moderated using ordinary water, tend to be simpler and cheaper to build than other types of nuclear reactor, and are well known to have excellent safety and stability characteristics. Due to these factors, they make up the vast majority of civil nuclear reactors in service throughout the world. LWRs can be subdivided into the following categories:Pressurized water reactors (PWRs), Boiling water reactors (BWRs)

- THE PRESSURIZED WATER REACTOR(PWR)

A pressurized-water reactor (PWR) uses ordinary light water as the reactor coolant and moderator in the state of high temperature and high pressure, not boiling in the reactor core (primary system: reactor coolant system) and sends the high-temperature and high-pressure water to steam generators (primary system) to generate steam with heat exchangers (steam system: secondary coolant system) for a turbine generator to generate electricity.

الفكرة العبرياتية لعمل المفاعل النووي هي واحدة في كل المفاعلات ولكن هناك نظامين مختلفين للتبريد حيث في النظام الاول يستخدم الماء المضغوط الذي يمكن ان ترتفع درجة حرارته إلى مئات الدرجات المئوية قبل ان يتحول الى بخار وبيستخدم الماء المضغوط كمصدر للحرارة لتحويل الماء إلى بخار في دائرة ثانوية أخرى منفصلة عن دائرة التبريد بينما 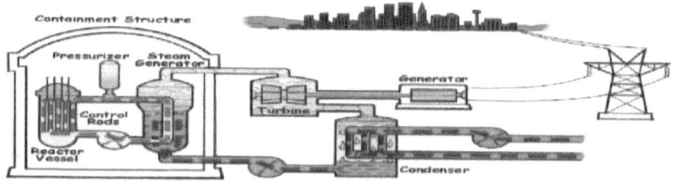 أنواع الاخرى من المفاعلات يتم ماء التبريد الذي ارتفعت درجة حرارته وتحول إلى بخار مباشرة لتحريك التوربينات وهنا تكون دائرة رئيسية واحدة كما هو موضح في المخططات التفصيلة التالية:

في الجزء الأيسر من مخطط المفاعل النووي يلاحظ الماء المضغوط الذي يستخدم في تبريد اليورانيوم والحرارة الناتج والتي بمنصها الماء المضغوط بمفدها لتحويل الماء إلى بخار يستخدم في تحريك التوربينات وتوليد الحركة المطلوبة لتوليد الطاقة الكهربية. لاحظ ان دائرة التبريد تختلف عن دائرة البخار.

هذا المخطط يوضح فكرة عمل المفاعل النووي المستخدم لتوليد الطاقة الكهربية ولكن هنا نجد ان الماء المستخدم في التبريد هو الذي يتحول إلى بخار ماء لتحريك التوربينات وتوليد الطاقة الكهربية. لاحظ هنا ان دائرة التبريد ودائرة البخار هي دائرة واحدة.

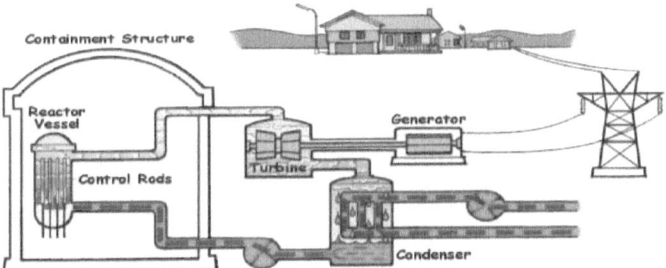

How solar power works, on-grid, off-grid and hybrid

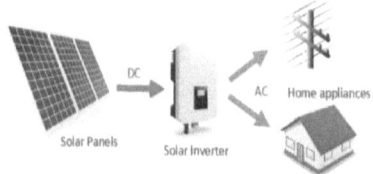

The main part of a solar electric system is the **solar panel**. There are various types of **solar panel** available in the market. **Solar panels** are also known as **photovoltaic solar panels**. Solar panel or solar module is basically an array of series and parallel connected **solar cells**. The potential difference developed across a solar cell is about 0.5 volt and hence desired number of such cells to be connected in series to achieve 14 to 18 volts to charge a standard battery of 12 volts. Solar panels are connected together to create a solar array. Multiple panels are connected together both in parallel and series to achieve higher current and higher voltage respectively.

Types of Solar Power Station

There are mainly two important types of solar power stations.
- Stand Alone or Off Grid type Solar Power Plant
- Grid Tie type Solar Power Plant

a. Components of Stand Alone Solar System

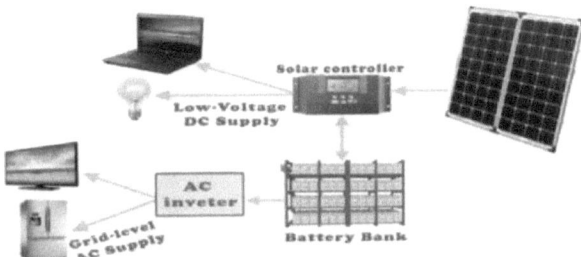

A basic block diagram of a stand-alone solar electric system is show above. Here the electric power produced in the solar panel is first supplied to the solar controller which in turn charges the battery bank or supplies directly to the low voltage DC equipments such as laptops and LED lighting system. Normally the battery is fed from solar controller but it can also feed the solar controller when there is insufficient supply of power from solar panel. In this way the supply is continued uniformly to the low voltage equipments which are directly connected to the solar controller. In this scheme the battery bank terminals are also connected across an inverter. The inverter converts the stored DC power of the battery bank to high voltage AC for running larger electrical equipments such as washing machines, larger televisions and kitchen appliances etc.

b. Components of Grid Tie Solar System

Grid tie solar systems are of two types one with single macro central inverter and other with multiple micro inverters. In the former type of solar system, the solar panels as well as grid supply are connected to a common central inverter called grid tie inverter as shown below

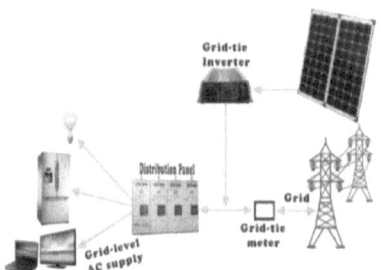

- The inverter here converts the DC of the solar panel to grid level AC and then feeds to the grid as well as the consumer's distribution panel depending upon the instantaneous demand of the systems. Here grid-tie inverter also monitors the power being supplied from the grid. If it finds any power cut in the grid, it actuates switching system of the solar system to disconnect it from the grid to ensure no solar electricity can be fed back to the grid during power cut. There is on energy meter connected in the main grid supply line to record the energy export to the grid and energy import from the grid.
As we already told there is another type of grid-tie system where multiple micro-inverters are used. Here one micro inverter is connected for each individual solar module. The basic block diagram of this system is very similar to previous one except the micro inverters are connected together to produce desired high AC voltage.

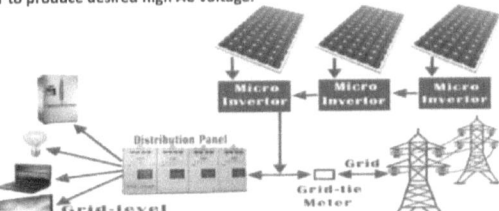

Part 2: Describe the main components and characteristics of thermal power plants.

- Identify the structure and the main components of thermal power plants.
- Describe various types of boilers and combustion process.
- List types of turbines, explain the efficiency of turbines, impulse turbines, reaction turbines, operation and maintenance, and speed regulation, and describe turbo generator.
- Explain the condenser cooling - water loop.
- Discuss thermal power plants and the impact on the environment.

2.1 Identifying the structure and the main components of thermal power plants and various types of boilers and combustion process. And explaining the condenser cooling - water loop

EEL 2023
Thermal Power Plants

THE RANKINE CYCLE

The Rankine cycle is standard for steam power plants that are built around the world. The basic Rankine cycle consists of four main phases:

- Steam Generator (boiler)

- Turbine

- Steam Condenser

- Pump

- **Process 1-2:** Water from the condenser at low pressure is pumped into the boiler at high pressure. This process is reversible adiabatic.
- **Process 2-3:** Water is converted into steam at constant pressure by the addition of heat in the boiler.
- **Process 3-4:** Reversible adiabatic expansion of steam in the steam turbine.
- **Process 4-1:** Constant pressure heat rejection in the condenser to convert condensate into water.

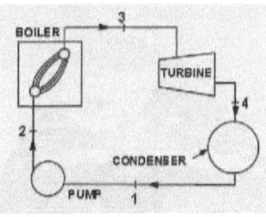

Basic Cycle

- The Rankine cycle is the fundamental operating cycle of all power plants where an operating fluid is continuously evaporated and condensed. The selection of operating fluid depends mainly on the available temperature range.

- The pressure-enthalpy (p-h) and temperature-entropy (T-s) diagrams of this cycle are given in Figure. The Rankine cycle operates in the following steps:

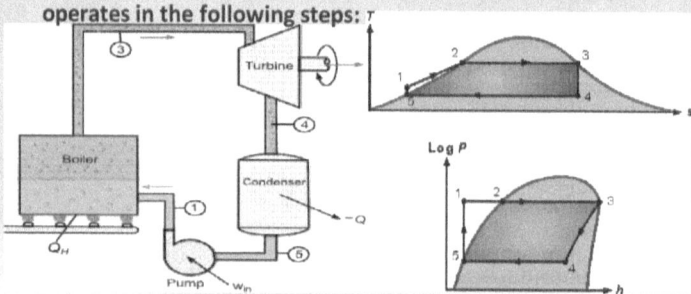

- **1-2-3 Isobaric Heat Transfer.** High pressure liquid enters the boiler from the feed pump (1) and is heated to the saturation temperature (2). Further addition of energy causes evaporation of the liquid until it is fully converted to saturated steam (3).

- **3-4 Isentropic Expansion.** The vapor is expanded in the turbine, thus producing work which may be converted to electricity. In practice, the expansion is limited by the temperature of the cooling medium and by the erosion of the turbine blades by liquid entrainment in the vapor stream as the process moves further into the two-phase region. Exit vapor qualities should be greater than 90%.

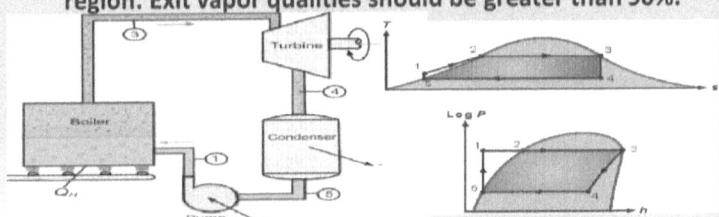

- **4-5 Isobaric Heat Rejection.** The vapor-liquid mixture leaving the turbine (4) is condensed at low pressure, usually in a surface condenser using cooling water. In well designed and maintained condensers, the pressure of the vapor is well below atmospheric pressure, approaching the saturation pressure of the operating fluid at the cooling water temperature.

- **5-1 Isentropic Compression.** The pressure of the condensate is raised in the feed pump. Because of the low specific volume of liquids, the pump work is relatively small and often neglected in thermodynamic calculations.

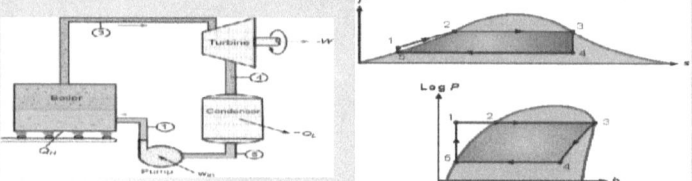

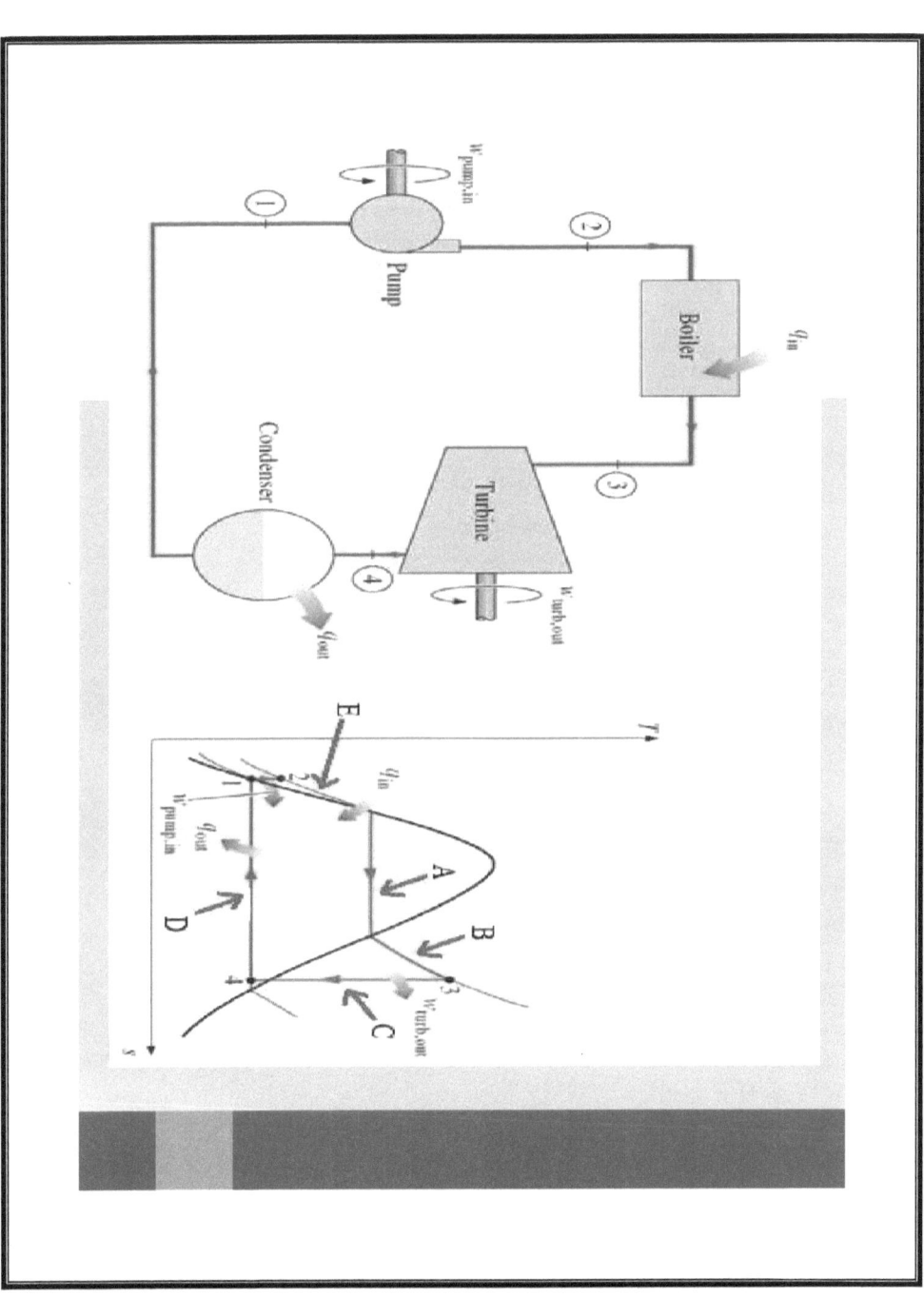

Figure 1 shows a temperature-entropy diagram or T-s diagram. Entropy is basically a measure of the amount of thermal energy not available to do work. The total area under the T-s diagram represents the work done during the complete cycle. The stages in the Rankine Cycle (fig.1) are :-

- E –the cold water is heated to boiling point

- A –heat, q_{in} is added to the water in the boiler at constant temperature converting it to steam

- B –more heat is added in a *superheater*, raising the temperature and *pressure* of the steam.

- C –the steam expands through the turbine giving up energy to turn the turbine and losing energy while its temperature drops.

- D –the steam is then cooled in a *condenser,* changing back to water and then *pumped* back into the boiler to continue the cycle

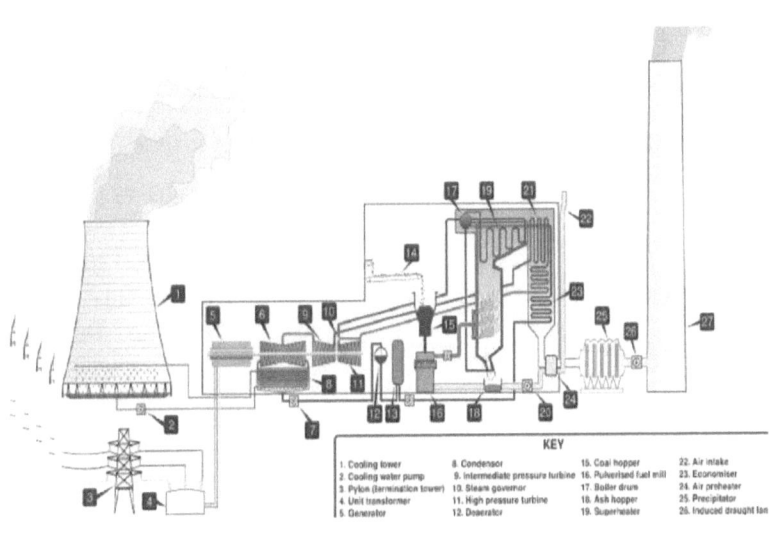

1. Cooling tower
2. Cooling water pump
3. Transmission line (3-phase)
4. Step-up transformer (3-phase)
5. Electrical generator (3-phase)
6. Low pressure steam turbine
7. Condensate pump
8. Surface condenser
9. Intermediate pressure steam turbine
10. Steam Control valve
11. High pressure steam turbine
12. Deaerator
13. Feedwater heater
14. Coal conveyor
15. Coal hopper
16. Coal pulverizer
17. Boiler steam drum
18. Bottom ash hopper
19. Superheater
20. Forced draught (draft) fan
21. Reheater
22. Combustin air intake
23. Economiser
24. Air preheater
25. Precipitator
26. Induced draught (draft) fan
27. Flue gas stack

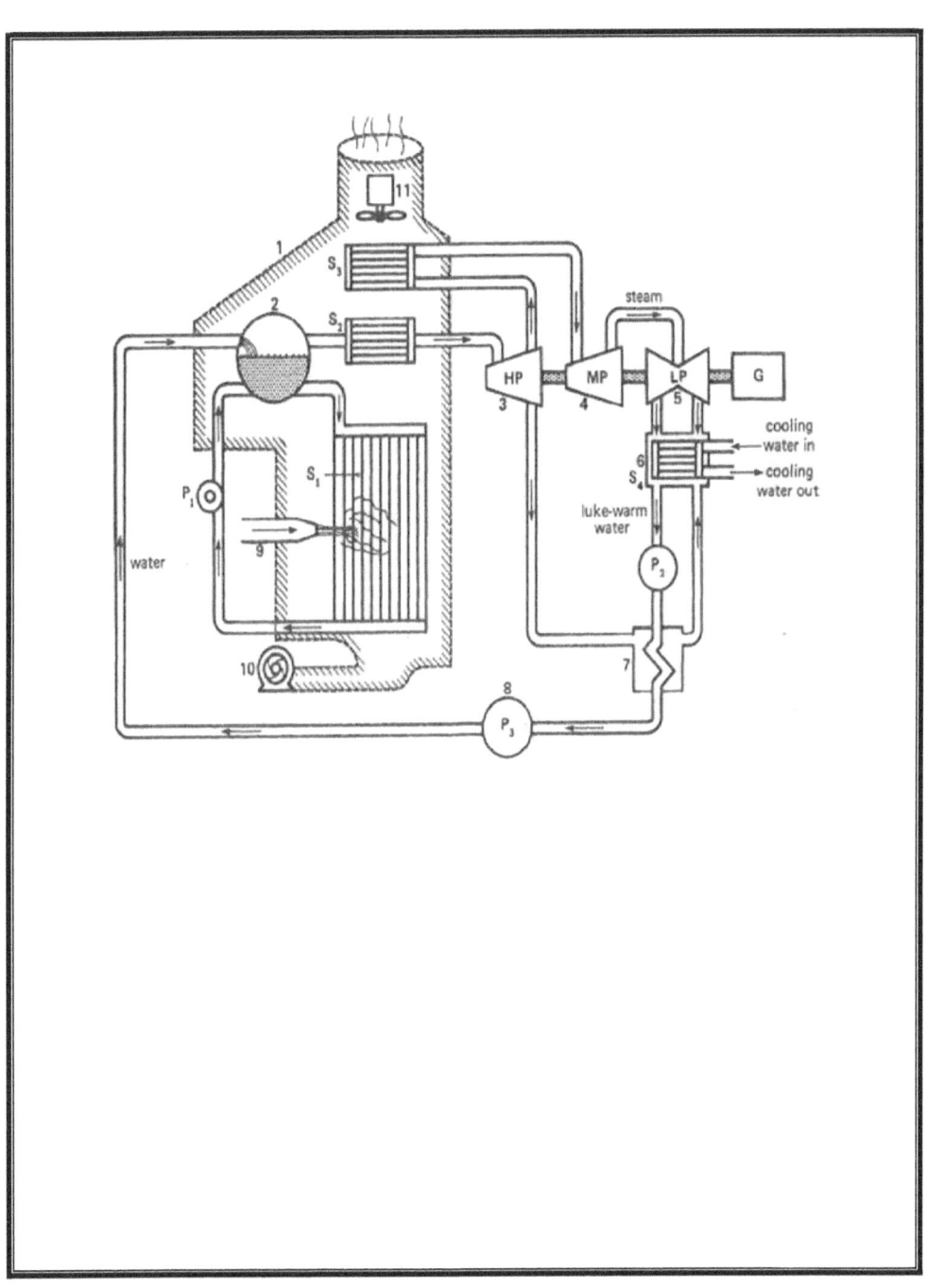

FOSSIL-FUELLED THERMAL POWER PLANT SYSTEM

Layout of thermal plant, an example of which is shown in Figures 1 & 2, can be easily understood by dividing the plant components into four main processes.

- 1-Coal and ash process.
- 2-Air and gas process.
- 3-Feed water and steam circuit.
- 4-Cooling water process.

Boiler operation

- The boiler is a rectangular furnace about 50 feet (15 m) on a side and 130 feet (40 m) tall..
- Pulverized coal is air-blown into the furnace through burners located at the four corners, or along one wall, or two opposite walls, and it is ignited to rapidly burn, forming a large fireball at the center. The thermal radiation of the fireball heats the water that circulates through the boiler tubes near the boiler perimeter. As the water in the boiler circulates it absorbs heat and changes into steam. It is separated from the water inside a drum at the top of the furnace. The saturated steam is introduced into superheat pendant tubes that hang in the hottest part of the combustion gases as they exit the furnace. Here the steam is superheated to 1,000 °F (540 °C) to prepare it for the turbine.

Boiler furnace and steam drum

The water enters the boiler through a section in the convection pass called the economizer. From the economizer it passes to the steam drum and from there it goes through downcomers to inlet headers at the bottom of the water walls. From these headers the water rises through the water walls of the furnace where some of it is turned into steam and the mixture of water and steam then re-enters the steam drum.

- When the water circulates out of the furnace and into the steam drum (17) at the top of the furnace, it is separated from the water. The steam is passed through a series of steam and water separators inside the steam drum.
- The *steam separators* remove water droplets from the steam and the cycle through the water walls is repeated. This steam is *saturated steam* which could harm the turbine blades.
- So, the saturated steam is pumped into the *super heater* (19) that hangs in the hottest part of the combustion gases as they exit the furnace.

Air path

- External fans are provided to give sufficient air for combustion. The Primary air fan takes air from the atmosphere and, first warms the air in the air preheater for better economy. Primary air then passes through the coal pulverizers, and carries the coal dust to the burners for injection into the furnace. The Secondary air fan takes air from the atmosphere and, first warms the air in the air preheater for better economy. Secondary air is mixed with the coal/primary air flow in the burners.
- The induced draft fan assists the FD fan by drawing out combustible gases from the furnace, maintaining a slightly negative pressure in the furnace to avoid leakage of combustion products from the boiler casing.

Coal and ash circuit

- Coal arrives at storage yard (14)
- Coal is directed in to the coal hopper (15)
- In case of *pulversing*, coal is pulverized (broken up) (16) and then goes to the fuel burners.
- Ash resulting from combustion of coal gets collected at the ash hopper (18) and is removed to ash storage yard by ash handling equipment.

The Pulveriser

The coal is put in the boiler after *pulverization*. A pulveriser is a device for grinding coal for combustion in a furnace in a power plant.

Feed water heating and deaeration

- The boiler feedwater used in the steam boiler is a means of transferring heat energy from the burning fuel to the mechanical energy of the spinning steam turbine. The total feed water consists of recirculated *condensate* water and purified *makeup water*.
- The feed water cycle begins with condensate water being pumped out of the condenser after traveling through the steam turbines.
- The water is pressurized in two stages, and flows through a series of six or seven intermediate feed water heaters, heated up at each point with steam extracted from an appropriate duct on the turbines and gaining temperature at each stage. Typically, in the middle of this series of feedwater heaters, and before the second stage of pressurization, the condensate plus the makeup water flows through a deaerator[9][10] that removes dissolved air from the water, further purifying and reducing its corrosiveness.

- The feed water system consists of re-circulated *condensate* water that is collected into the surface condenser (8) and pumped out by the condensate pump (7), and purified *makeup water*.
- The feed water cycle begins with condensate water being pumped out of the *condenser* (7) after coming out of the steam turbines in to the *deaerator* (12) whose function is to remove dissolved air from the water, further purifying and reducing its *corrosivity*.

- When the water circulates out of the furnace and into the steam drum (17) at the top of the furnace, it is separated from the water. The steam is passed through a series of steam and water separators inside the steam drum.
- The *steam separators* remove water droplets from the steam and the cycle through the water walls is repeated. This steam is *saturated steam* which could harm the turbine blades.
- So, the saturated steam is pumped into the *super heater* (19) that hangs in the hottest part of the combustion gases as they exit the furnace.

Air path

- External fans are provided to give sufficient air for combustion. The Primary air fan takes air from the atmosphere and, first warms the air in the air preheater for better economy. Primary air then passes through the coal pulverizers, and carries the coal dust to the burners for injection into the furnace. The Secondary air fan takes air from the atmosphere and, first warms the air in the air preheater for better economy. Secondary air is mixed with the coal/primary air flow in the burners.

- The induced draft fan assists the FD fan by drawing out combustible gases from the furnace, maintaining a slightly negative pressure in the furnace to avoid leakage of combustion products from the boiler casing.

Coal and ash circuit

- Coal arrives at storage yard (14)
- Coal is directed in to the coal hopper (15)
- In case of *pulversing*, coal is pulverized (broken up) (16) and then goes to the fuel burners.
- Ash resulting from combustion of coal gets collected at the ash hopper (18) and is removed to ash storage yard by ash handling equipment.

The Pulveriser

The coal is put in the boiler after *pulverization*. A pulveriser is a device for grinding coal for combustion in a furnace in a power plant.

Condenser

- Steam from the exhaust of the turbine is taken into the condenser so that it is turned into water to allow it to be pumped. Figure shows a typical water-cooled condenser
- . The condenser is made of a *shell* to contain the inlet steam and tubes in which cooling water is circulated.
- The exhaust steam from the low pressure stage of the turbine enters the shell where it is cooled and converted to water by flowing over the tubes as shown in Figure.
- The condenser generally uses either circulating cooling water from a *cooling tower* to reject waste heat to the atmosphere, or once-through water from a river, lake, or ocean.

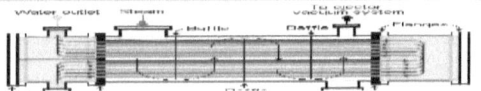

Cooling Water Circuit:

The *condenser* (8) requires cooling water to condense the exhaust steam. The water is cooled in cooling towers (1) or in cooling ponds and reused again and again. Some make up cooling water is added in the circuit.

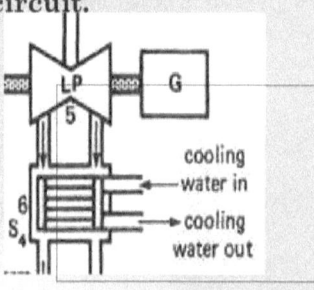
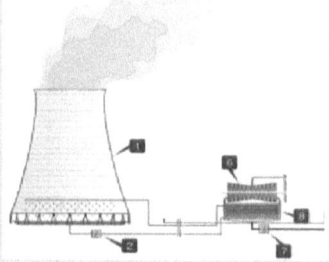

Re-heater (21)

Power plant furnaces may have a reheater section containing tubes heated by hot flue gases outside the tubes.

The re-heater section contains tubes heated by hot flue gases outside the tubes. Exhaust steam from the high pressure turbine is rerouted to go inside the re heater tubes to pick up more energy to drive intermediate or lower pressure turbines.

Superheater

Fossil fuel power plants often have a <u>superheater</u> section in the steam generating furnace [1] The steam passes through drying equipment inside the steam drum on to the superheater, a set of tubes in the furnace. Here the steam picks up more energy from hot flue gases outside the tubing, and its temperature is now superheated above the saturation temperature. The superheated steam is then piped through the main steam lines to the valves before the high-pressure turbine

10- Boiler Types

A number of different boiler types have been developed to suit the various steam applications. The two basic types of boilers are:

a) Firetube: In firetube boilers, the combustion gases pass inside boiler tubes, and heat is transferred to water on the shell side.

b) Watertub: The *Scotch Marine boilers* are the most common type of industrial firetube boilers. Scotch marine boilers are typically cylindrical shells with horizontal tubes configured such that the exhaust gases pass through these tubes, transferring energy to boiler water on the shell

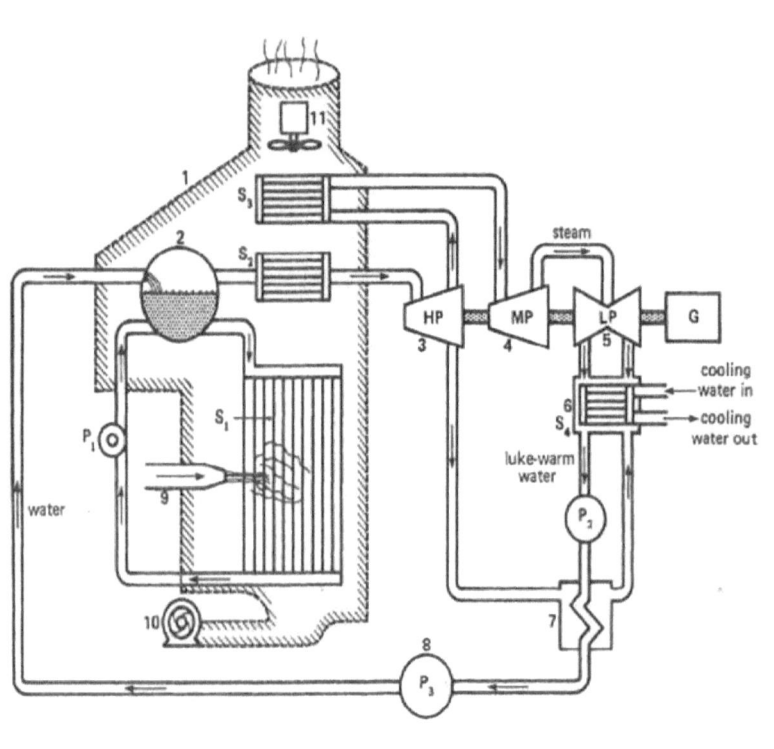

a) Fire Tube Boiler

As it indicated from the name, the fire tube boiler consists of numbers of tubes through which hot gasses are passed. These hot gas tubes are immersed into water, in a closed vessel. Actually in fire tube boiler one closed vessel or shell contains water, through which hot tubes are passed. These fire tubes or hot gas tubes heated up the water and convert the water into steam and the steam remains in same vessel. As the water and steam both are in same vessel a fire tube boiler cannot produce steam at very high pressure. Generally it can produce maximum 17.5 kg/cm^2 and with a capacity of 9 Metric Ton of steam per hour.

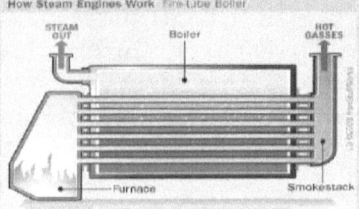

In the early 1950s, the UK Ministry of Fuel and Power sponsored research into improving boiler plant. The outcome of this research was the packaged boiler, Mostly, these boilers were designed to use oil rather than the coal.

FIRETUBE BOILER

 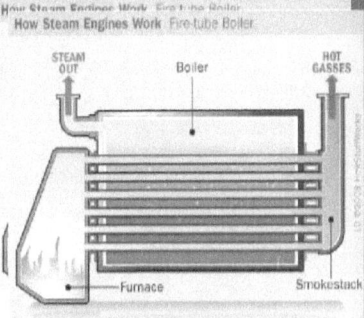

Figure 6 – A firetubeboiler

b) Water Tube Boiler

A water tube boiler is such kind of boiler where the water is heated inside tubes and the hot gasses surround them. This is the basic definition of water tube boiler. Actually this boiler is just opposite of fire tube boiler where hot gasses are passed through tubes which are surrounded by water.

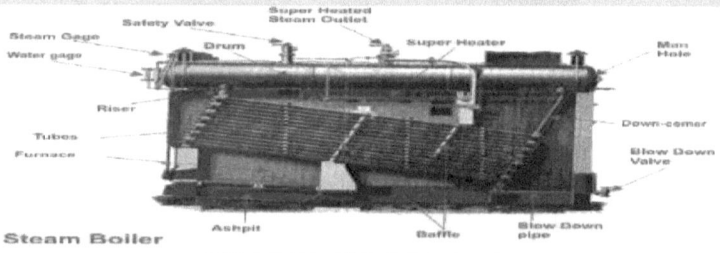

Steam Boiler

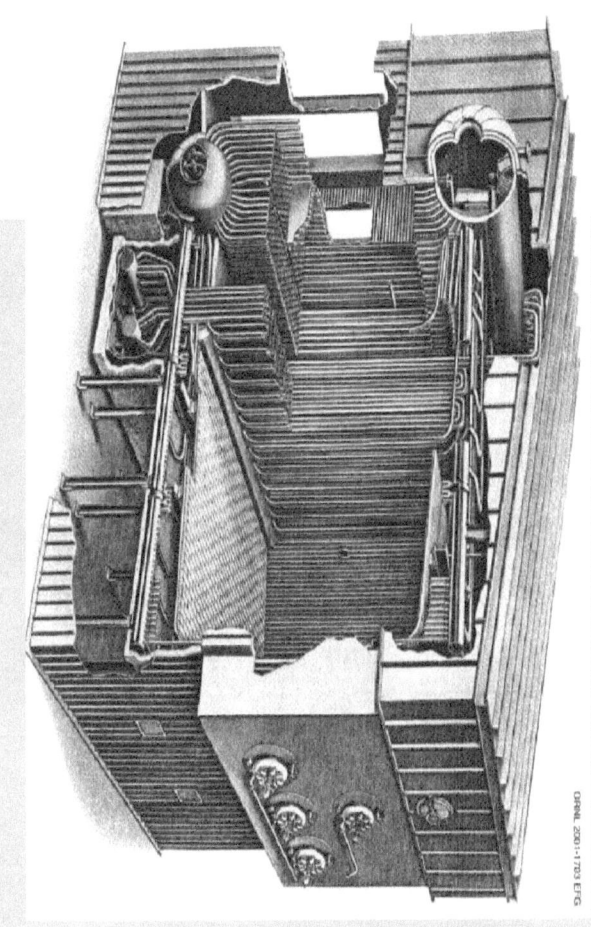

Scott Marine boilers

The *Scotch Marine boilers* are the most common type of industrial firetube boilers. Scotch marine boilers are typically cylindrical shells with horizontal tubes configured such that the exhaust gases pass through these tubes, transferring energy to boiler water on the shell side.

- *Advantages of the Scott Marine boiler are:*
  - Low initial cost.
  - High efficiency and durability.
  - Contains relatively large amount of water, which enables them to respond to load changes with relatively little change in pressure.

- *Disadvantages of the Scott Marine boilers:*
  - Since the boiler holds a large water mass, it requires more time to initiate steaming and more time to accommodate changes in steam pressure.
  - Scotch Marine boilers generate steam on the shell side, which has a large surface area, limiting the amount of pressure they can generate.

APPLICATION OF THE SCOTT MARINE BOILERS

- Scotch marine boilers are not used here pressures above 300 psig are required.
- Firetube boilers are often characterized by their number of passes, referring to the number of times the combustion (or flue) gases flow the length of the pressure vessel as they transfer heat to the water.
- Each pass sends the flue gases through the tubes in the opposite direction. To make another pass, the gases turn 180° and pass back through the shell.

Advantages of watertube boilers:

- Since tubes can withstand higher internal pressure than the large chamber shell in a firetube, watertube boilers are used where high steam pressures (3,000 psi, sometimes higher) are required.
- Watertube boilers are also capable of high efficiencies and can generate saturated or superheated steam.
- These boilers are attractive in applications that require dry, high-pressure, high-energy steam, including steam turbine power generation.

ADVANTAGES AND DISADVANTAGES OF THERMAL POWER PLANTS

Advantages

- They can be located very conveniently near the load centers.
- Does not require shielding like required in nuclear power plants.
- Unlike nuclear power plants whose power production method is difficult, for thermal power plants it is easy if compared.
- Transmission costs are reduced as they can be set up near the industry.
- The portion of steam generated can be used as process steam in different industries such as water desalination.
- Steam engines and turbines can work under 25% of overload capacity.
- Able to respond to changing loads without difficulty.

Disadvantages:

- Large amounts of water are required.
- Great difficulties experienced in coal handling and disposal of ash.
- Takes long time to be erected and put into action.
- Maintenance and operating costs are high.
- With increase in pressure and temperature, the cost of plant increases.
- Troubles from smoke and heat from the plant

2.2. Listing types of turbines, explain the efficiency of turbines, impulse turbines, reaction turbines, operation and maintenance, and speed regulation, and describe turbo generator.

STEAM TURBINES

- A steam turbine is a mechanical device that extracts *thermal energy* from pressurized steam, and converts it into rotary motion. Its modern equivalent was invented by *Sir Charles Parsons* in 1884.
- By far the most widely used and most powerful turbines are those driven by steam. Until the 1960s essentially all steam used in turbine cycles was raised in boilers burning fossil fuels (coal, oil, and gas) or, in minor quantities, certain waste products.
- However, modern turbine technology includes nuclear steam plants as well as production of steam supplies from other sources.

Principle of Operation of a Steam Turbine

- A steam turbine uses steam to rotate its blades. The rotary motion of the blades is used to rotate the armature of the generator, and the movement of the armature in a magnetic field results in the production of a current (electricity) in the armature!
- Heat energy from a thermal power plant or a nuclear power plant is used to boil water, and convert it into steam at high pressure. This high pressure steam is directed to the turbine blades thus causing the blades to rotate!
- Figures 8 & 9 shows the operation of the steam turbine.

STEAM TURBINE OPERATION

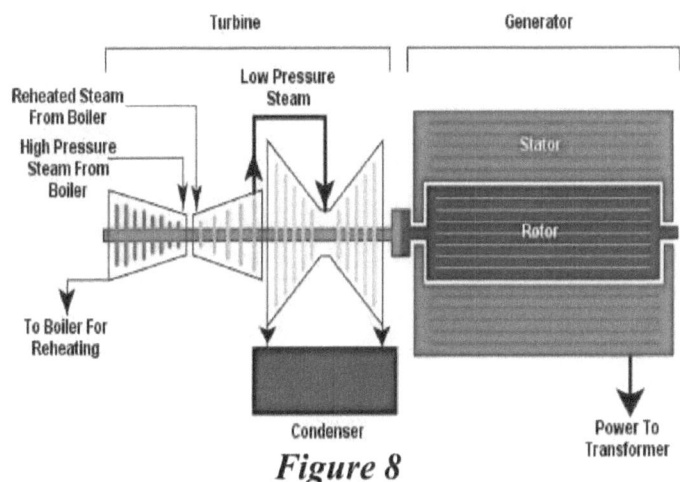

Figure 8

STEAM TURBINE OPERATION

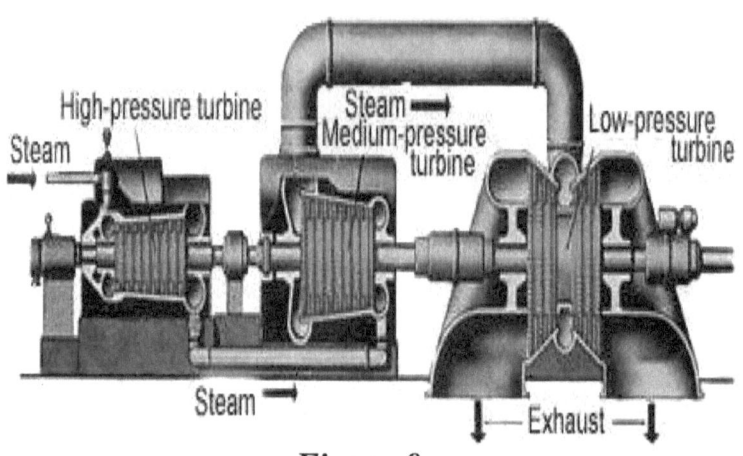

Figure 9

STEAM TURBINE CONSTRUCTION

Figure 10

STEAM TURBINE DESIGN - TURBINE BLADES

- Steam turbines- the blades are designed in such a way as to produce maximum rotational energy by directing the flow of the steam along its surface.
- The blades of a steam turbine are designed to behave like nozzles, thus effectively tapping both the impulse and reaction force of the steam for higher efficiency.
- Nozzle design itself is a complex process, and the nozzle shaped blade of the turbine is probably one of the most important parts in its construction.
- The blades are made at specific angles in order to incorporate the net flow of steam over it in its favour. The blades may be of stationary or fixed and rotary or moving types.

TORQUE PRODUCTION

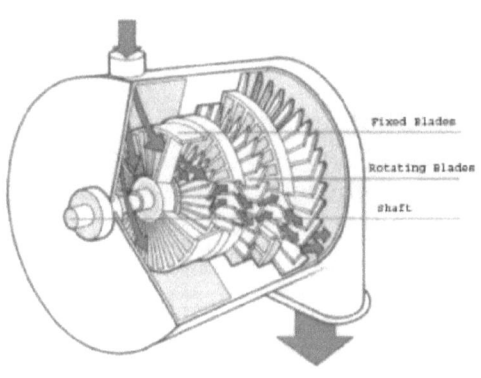

Steam is injected At the low pressure (LP) end.
It is directed by the stationary blades to pass through the spaces of the rotating blades. This creates a large torque which rotates the entire turbine shaft.

Figure 11

Impulse and reaction turbines

In an impulse turbine, a fast-moving fluid is fired through a narrow nozzle at the turbine blades to make them spin around. The blades of an impulse turbine are usually bucket-shaped so they catch the fluid and direct it off at an angle or sometimes even back the way it came (because that gives the most efficient transfer of energy from the fluid to the turbine). In an impulse turbine, the fluid is forced to hit the turbine at high speed.

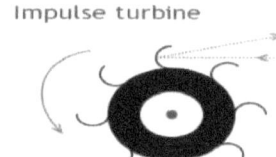

Impulse turbine

In a reaction turbine, the blades sit in a much larger volume of fluid and turn around as the fluid flows past them. A reaction turbine doesn't change the direction of the fluid flow as drastically as an impulse turbine: it simply spins as the fluid pushes through and past its blades. Wind turbines are perhaps the most familiar examples of reaction turbines.

Reaction turbine

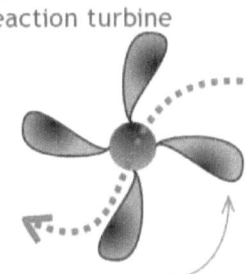

IMPULSE TURBINES

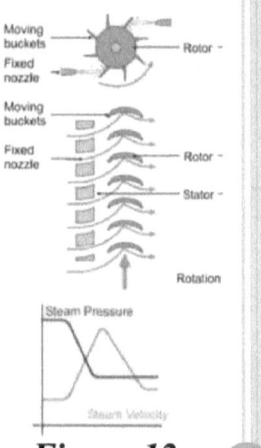

- An impulse turbine has fixed nozzles that direct the steam flow into high speed jets.

- These jets contain significant kinetic energy, which the rotor blades, shaped like buckets, convert into shaft rotation as the steam jet changes direction.

- A pressure drop occurs across only the stationary blades, with a net increase in steam velocity across the stage.

Figure 12

IMPULSE TURBINE OPERATION

- As the steam flows through the nozzle its pressure falls from inlet pressure to the exit pressure (atmospheric pressure, or more usually, the condenser vacuum).
- Due to this higher ratio of expansion of steam in the nozzle the steam leaves the nozzle with a very high velocity. The steam leaving the moving blades has a large portion of the maximum velocity of the steam when leaving the nozzle.
- The loss of energy due to this higher exit velocity is commonly called the *carry over velocity* or *leaving loss*.
- Figure 14 to 17 shows photos of steam turbines.

IMPULSE TURBINE BLADES

Rotor Blades showing impulse arrangement

Figure 14

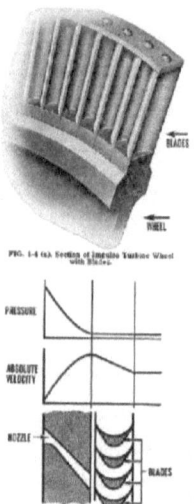

STEAM TURBINE ROTOR

A high-pressure rotor at a closer look. It has 20 steps. Note a massive steel turbine body consisted of two parts and pins with which

Figure 15

STEAM TURBINE ROTOR

Figure 16

NOZZLE ARRAYS

Figure 17

THERMAL POWER PLANT EFFICIENCY

- The efficiency of thermal generating stations is always low because of the inherent low efficiency of the turbines.
- The maximum efficiency of any machine that converts heat energy into mechanical energy is given by the equation:

$$\eta = (1 - \frac{T_2}{T_1}) \times 100$$

where,
η is the efficiency
T_1 = Temperature of the gas entering the turbine (in Kelvin)
T_2 = Temperature of the gas leaving the turbine (in Kelvin)

THERMAL POWER PLANT EFFICIENCY

- In most thermal generating stations the gas is steam. In order to obtain a high efficiency, the quotient T_2/T_1 should be as small as possible.

- However, temperature T_2 cannot be lower the ambient temperature, which is usually about $20°C$. As a result, T_2 cannot be less than

$$T_2 = 20 + 273 = 293\ K$$

THERMAL POWER PLANT EFFICIENCY

- This means that to obtain high efficiency, T_1 should be as high as possible. The problem is that we cannot use temperatures above those that steel and other metals can safely withstand, bearing in mind the corresponding high steam pressures.

- It turns out that the highest feasible temperature T_1 is about $550°C$. As a result,

$$T_1 = 550° + 273° = 823\,K$$

- It follows that the maximum possible efficiency of a turbine driven by steam that enters at $823\,K$ and exits at $293\,K$ is

$$\eta = \left(1 - \frac{293}{823}\right) \times 100 = 64.4\%$$

THERMAL POWER PLANT EFFICIENCY

- Due to other losses, some of the most efficient steam turbines have efficiencies of 45%. This means that 65% of the thermal energy is lost during the thermal-to-mechanical conversion process.
- The enormous loss of heat and how to dispose of it represents one of the major aspects of a thermal generating station.

THERMAL POWER PLANT EFFICIENCY

Example:

Find the thermal power plant efficiency if the input steam temperature to the turbine is 700 K, and the output steam temperature from the turbine is 77 °C?

- Thermal Power Station Simulation
- http://www.youtube.com/watch?v=xokHLFE96h8
- Steam Turbine:
- http://www.youtube.com/watch?v=MulWTBx3szc
- http://www.learnengineering.org/2013/02/working-of-steam-turbine.html
- Steam Turbine Sulzer
- http://www.youtube.com/watch?v=ZOkbw5uVQww&feature=related
- Ontario Thermal Plant
- http://www.youtube.com/watch?v=SeXG8K5_UvU
- Steam Turbine (+HRSG)
- http://www.youtube.com/watch?NR=1&v=1bl1Q3V_79I&feature=fvwp
- Shell & Tube Heat Exchanger
- http://www.youtube.com/watch?v=hxhB3k0vh2g

2.3 Discuss thermal power plants and the impact on the environment.

EEL 2023

POWER GENERATION AND TRANSMISSION

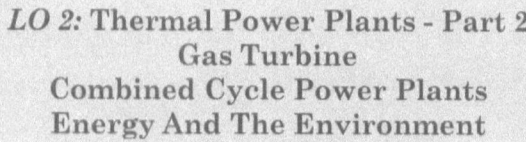

LO 2: Thermal Power Plants - Part 2
Gas Turbine
Combined Cycle Power Plants
Energy And The Environment

GAS TURBINES

- A gas turbine extracts energy from the flow of hot gas produced by the combustion of gas or fuel oil in a stream of compressed air.
- It has an upstream air compressor, mechanically coupled to a downstream turbine, and a combustion chamber in between.

Gas Turbines

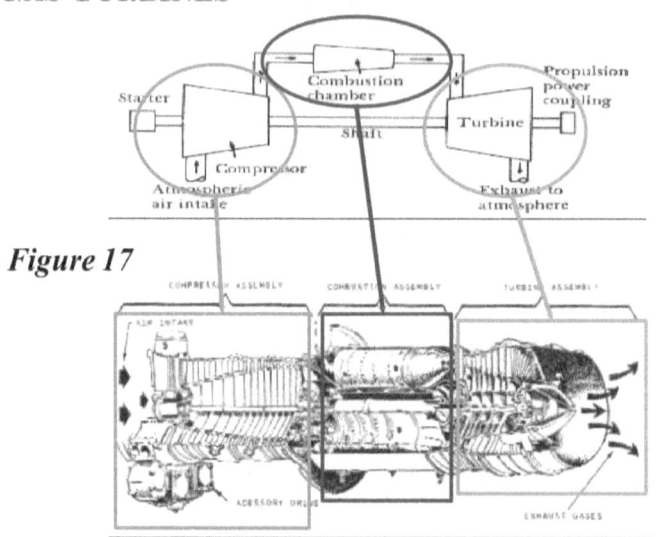

Figure 17

The Gas Turbine Cycle (1)

- Intake phase: Outside air is drawn into the engine by the action of the compressor. Pressure, temperature and volume remain the same through the intake phase.
- Compression phase: Intake air is mechanically compressed. Pressure and temperature increase with a corresponding decrease in volume. Mechanical energy driving the compressor is converted to kinetic energy in the form of compressed air.
- Combustion phase: Fuel is sprayed into the combustor and burned, converting the chemical energy to thermal energy in the form of a hot expanding gas. Volume and temperature greatly increase while pressure remains constant through the combustor.

THE GAS TURBINE CYCLE (2)

- Expansion phase: Thermal energy is converted to mechanical energy as the hot expanding gases from the combustor turn the turbine rotor. Pressure and temperature decrease while volume increases through the expansion phase.
- Exhaust phase: Hot exhaust gases are ducted through exhaust ducts to the atmosphere. Pressure, temperature and volume remain the same through the exhaust phase.
- Combustion Chamber: The combustion chamber mixes fuel with compressed air and ignites the mixture.
- Turbine: Hot gases from the combustion chamber are used to do work by driving loads such as electrical generators.

GAS TURBINE OPERATION

- Energy is released when compressed air is mixed with fuel and ignited in the combustion chamber as shown in.
- The gases are passed through a nozzle onto the turbine blades, generating thrust by accelerating the hot exhaust gases by expansion back to atmospheric pressure.
- Energy is extracted in the form of shaft power, and used to power aircraft, trains, ships & electrical generators.

GAS TURBINE OPERATION

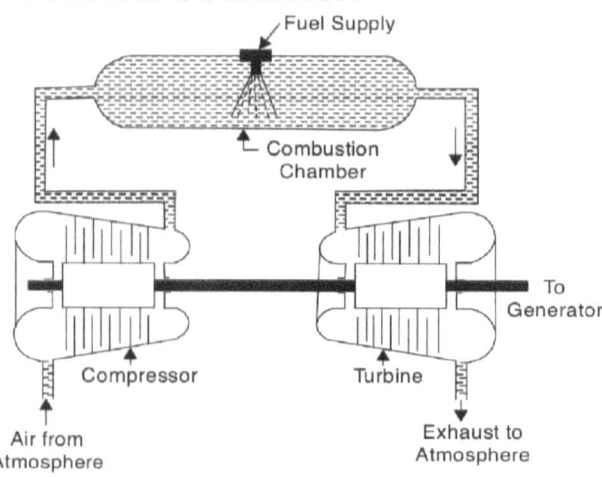

Figure 18 – Gas Turbine Operation

GAS TURBINE OPERATION

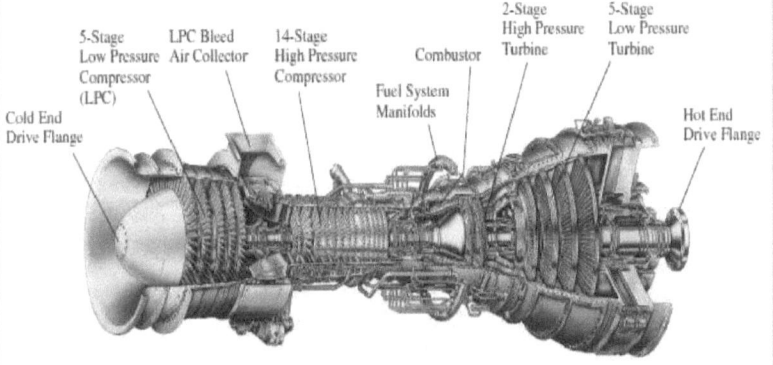

Figure 19: Cutaway view of a gas turbine

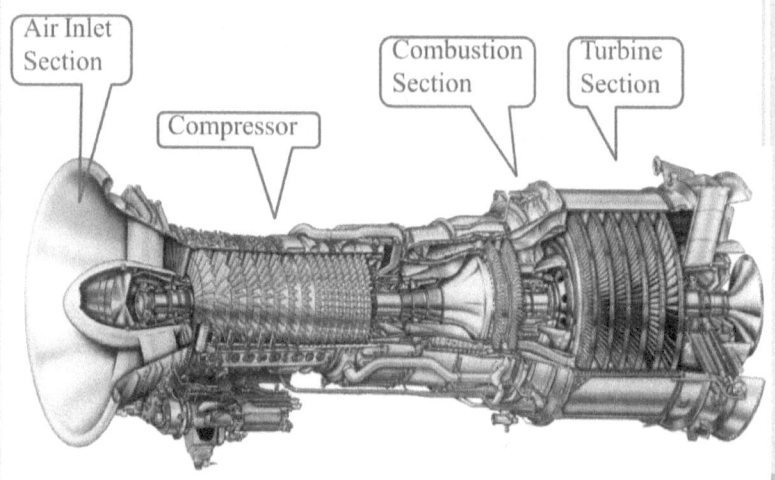

General Electric LM2500 Gas Turbine

GAS TURBINE MAIN COMPONENT

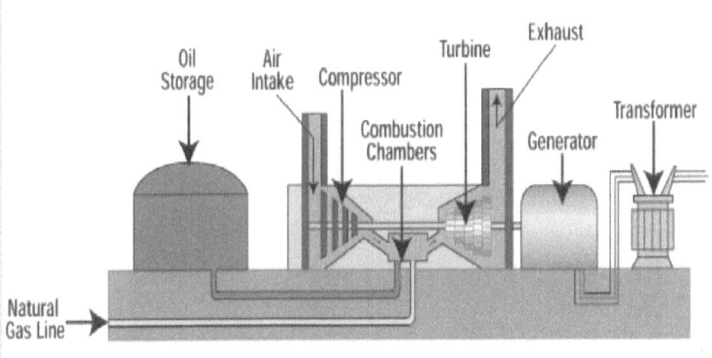

BRAYTON CYCLE – GAS TURBINE CYCLE

The **Brayton cycle** or **Joule Cycle** is made up of four internally reversible processes:

1-2 Compression

2-3 Heat addition

3-4 Expansion

4-1 Heat rejection

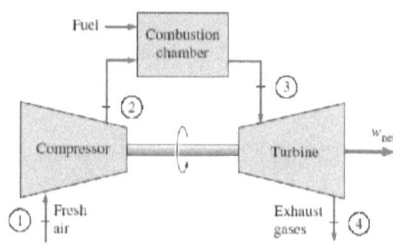

Entropy

Entropy on Thermodynamics. (on a macroscopic scale) a function of thermodynamic variables, as temperature, pressure, or composition, that is a measure of the energy that is not available for work during a thermodynamic process. A closed system evolves toward a state of maximum entropy.

Brayton Cycle – Gas Turbine Cycle

The Brayton cycle is made up of four internally reversible processes:

1-2 Isentropic (constant entropy) compression (in a compressor)

2-3 Constant-pressure heat addition

3-4 Isentropic expansion (in a turbine)

4-1 Constant-pressure heat rejection

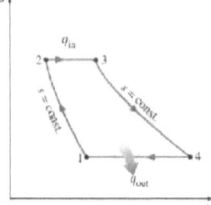

(b) P-V diagram
Pressure vs. Volume

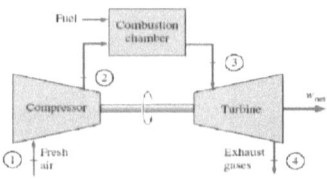

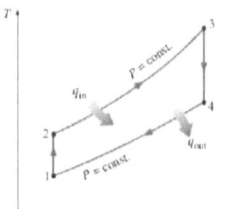

(a) T-s diagram
Temperature vs. Entropy

OVERALL THERMAL EFFICIENCY

- Useful Work = Energy released in turbine minus energy absorbed by compressor.

- The compressor requires typically approximately 30% - 50% of the energy released by the turbine.

$$Overall\ Thermal\ Efficiency = \frac{Useful\ Work}{Fuel\ Chemical\ Energy} * 100\%$$

- Typical overall thermal efficiency of a combustion turbine is 20% - 40%.

OVERALL THERMAL EFFICIENCY

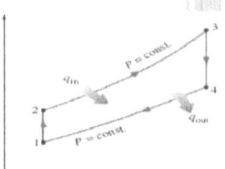

$$Overall\ Thermal\ Efficiency = \frac{Useful\ Work}{Fuel\ Chemical\ Energy}$$

$$\eta = \frac{P_{net}}{Q_{in}}$$

$$\eta = 1 - \frac{T_4 - T_1}{T_3 - T_2}$$

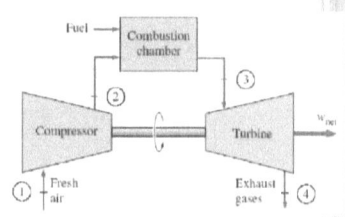

$$\eta = 1 - r^{-0.286}$$

where r is the compression ratio, and temperatures are in Kelvin.

Example:

A gas turbine has a pressure ratio of 6:1. The inlet temperature to the compressor is 10°C. The outlet temperature from the compressor is 199.4°. The inlet temperature to the turbine is 950°C. Find:

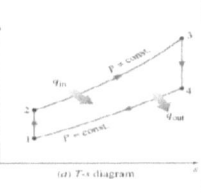
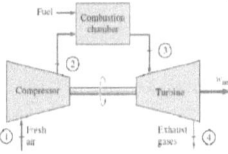

a) The gas turbine cycle thermal efficiency.

$$\eta = 1 - r^{-0.286}$$
$$\eta = 40\%$$

b) The outlet temperature from the turbine.

$$\eta = 1 - \frac{T_4 - T_1}{T_3 - T_2}$$

$T_1 = 10+273 = 283$ K
$T_2 = 199.4+273 = 472.4$ K
$T_3 = 950+273 = 1223$ K

→ $T_4 = 733.36$ K $= 460.36$ °C.

c) The net output power if the inlet heat power is 150.8 MW.
$Q_{in} = 150.8$ MW

$$\eta = \frac{P_{net}}{Q_{in}}$$

$P_{net} = 60.32$ MW

EXAMPLE 2

Calculate the thermal efficiency of a gas turbine if the net output power is 42 MW when the inlet heat energy is 120 MW.

GAS TURBINE CONSTRUCTION

Questions : a. Where is the compressor?
b. Where is the power turbine?   c. Where are the combustion chambers ?

Combustion Turbine

GAS TURBINE INLET FILTER SYSTEM

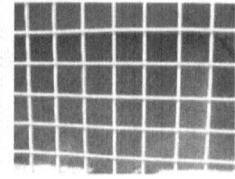

GAS TURBINE INLET FILTER SYSTEM

Inlet Guide Vanes

Inlet Guide Vanes

Gas Turbine

GAS TURBINE CUT AWAY SIDE VIEW

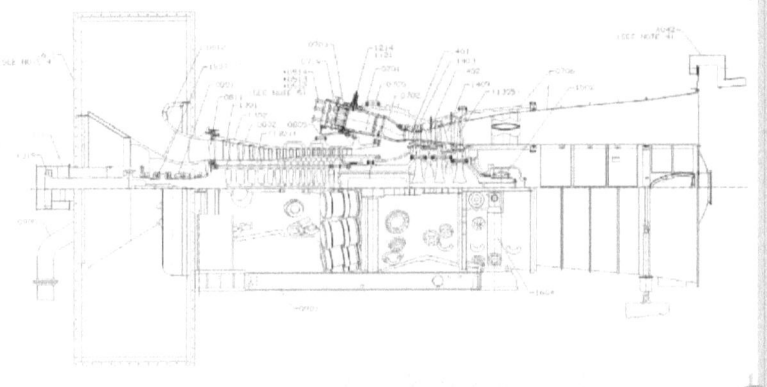

GAS TURBINE COMBUSTOR ARRANGEMENT

FRAME 5 GT

Typical Simple Cycle CT Plant Components

- Prime Mover (Combustion Turbine)
- Fuel Supply & Preparation
- Emissions Control Equipment
- Generator
- Electrical Switchgear
- Generator Step Up Transformer
- Starting System (Combustion Turbines)
- Auxiliary Cooling
- Fire Protection
- Lubrication System

Typical CT Plant Components

Lube Oil System

Step Up Transformer

Generator

Switchgear / MCC

Starting Motor

Fire Protection

GAS TURBINE ADVANTAGES

- Relatively low cost to build
- Relatively Short Build time
- Packaged units – low commissioning period
- Fast starting (~5 minutes – ideal for *peak demand*)
- Very large power-to-weight ratio
- Long working life
- Can run on gas or liquid fuel (oil, kerosene, etc ...)

VIDEO TIME

- http://www.youtube.com/watch?v=r9q80sSHxKM&feature=endscreen

- http://www.youtube.com/watch?v=c12Gh8BN0Io&feature=endscreen

COMBINED CYCLE POWER PLANTS

- A *combined* cycle gas turbine power plant is essentially an electrical power plant in which a *gas turbine* and a *steam turbine* are used in combination to achieve greater efficiency than would be possible independently.
- The gas turbine drives an *electrical generator* while the gas turbine exhaust is used to produce steam in a *heat exchanger* (called a *Heat Recovery Steam Generator, HRSG*).
- The steam is then supplied to a steam turbine whose output provides the means to generate *more* electricity.
- If the steam is used for heating (heating buildings, water desalination, etc ...) then the plant would be referred to as a *cogeneration* plant.

COMBINED CYCLE POWER PLANT
- GAS TURBINES PLUS STEAM TURBINES

- Gas Turbine Exhaust used as the heat source for the Steam Turbine cycle
- Utilizes the major efficiency loss from the GT cycle
- Advantages:
    - Relatively short cycle to design, construct & commission
    - Higher overall efficiency
    - Good cycling capabilities
    - Fast starting and loading
    - Lower installed costs
    - No issues with ash disposal or coal storage
- Disadvantages (applicable for both SCGT and CCGT)
    - High fuel costs
    - Uncertain long term fuel source
    - Output dependent on ambient (air) temperature

COMBINED CYCLE POWER PLANT
COMBINING THE BRAYTON AND RANKINE CYCLES

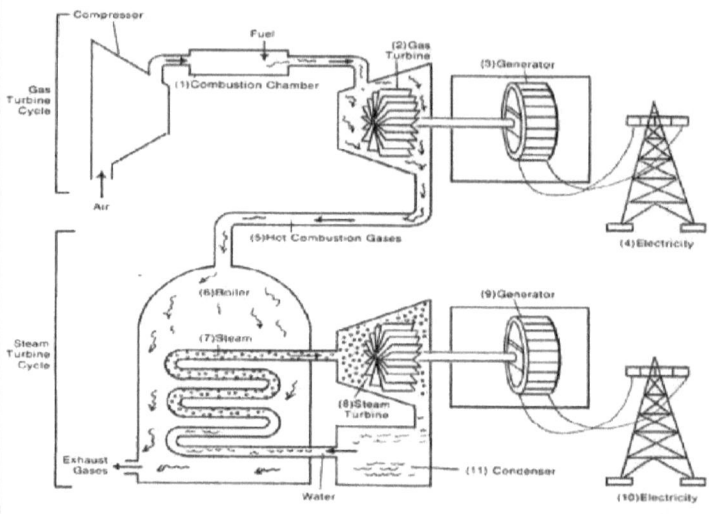

OPERATION OF COMBINED CYCLE

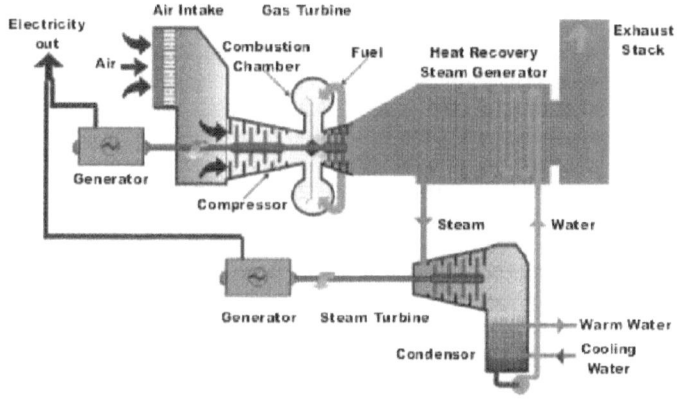

COMBINED CYCLE PLANT LAYOUT

- The steam cycle part uses a heat exchanger instead of a boiler heated by fossil fuel.
- The hot exhaust gases from the GT are used to produce the steam for the steam turbine.
- The efficiency of a GT alone (*simple cycle*) is about 35%.
- The efficiency of *combined cycle* is about 58-60%.

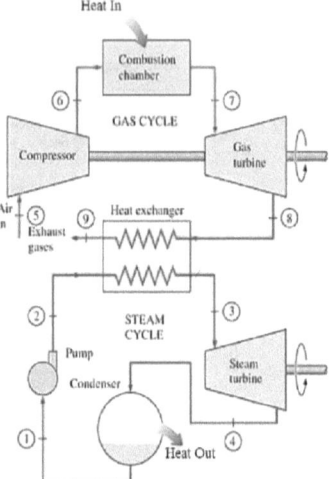

COMBINED CYCLE POWER PLANT

Heat Recovery Steam Generator (HRSG)

COMBINED CYCLES TODAY

- **Plant Efficiency ~ 58-60 percent**
  - Biggest losses are mechanical input to the compressor and heat in the exhaust
- **Steam Turbine output**
  - Typically 50% of the gas turbine output
  - More with duct-firing (burning extra fuel in the heat exchanger)
- **Net Plant Output** (Using General Electric Industrial gas turbines)
  - up to 750 MW for 3 on 1 configuration
  - Up to 520 MW for 2 on 1 configuration
- **Construction time about 24 months**
- **Engineering duration about 12 months**
- **Capital Cost ($900-$1100/kW)**

Combined heat and power (Cogeneration)

Cogeneration is also called as combined heat and power or combine heat and power. As it name indicates cogeneration works on concept of producing two different form of energy by using one single source of fuel. Out of these two forms one must be heat or thermal energy and other one is either electrical or mechanical energy.
Cogeneration is the most optimum, reliable, clean and efficient way of utilizing fuel. The fuel used may be natural gas, oil, diesel, propane, wood, bassage, coal etc. It works on very simple principle i.e the fuel is used to generate electricity and this electricity produces heat and this heat is used to boil water to produce steam, for space heating and even in cooling buildings.

COGENERATION

- Cogeneration is the simultaneous production of various forms of energy from one power source (normally associated with heat and power-mechanical or electrical).
- The engine produces primary electrical power whereas thermal energy in the exhaust gases is converted in the heat recovery boiler HRB into steam. This process steam could be used in another application such as heating the sea water in the desalination units.

- A Cogeneration Plant
  - Power generation facility that also provides thermal energy (steam) to a thermal host.
- Typical thermal hosts
  - water desalination plant
  - paper mills,
  - chemical plants,
  - refineries, etc...
  - potentially any user that uses large quantities of steam on a continuous basis.

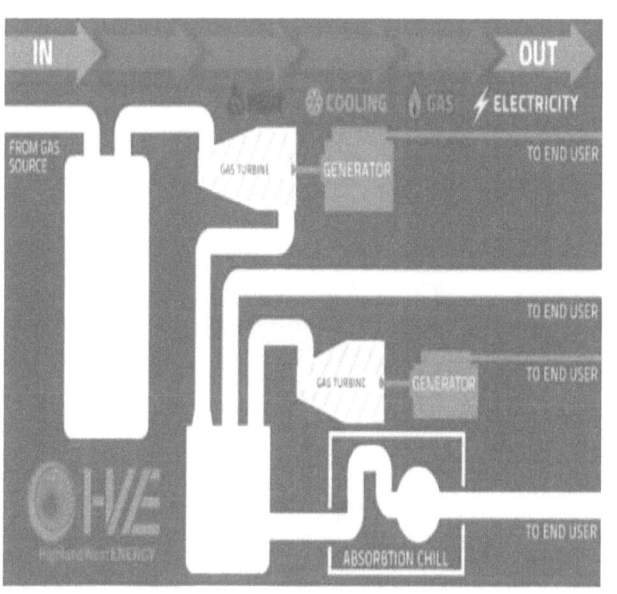

Power and Desalination Cogeneration

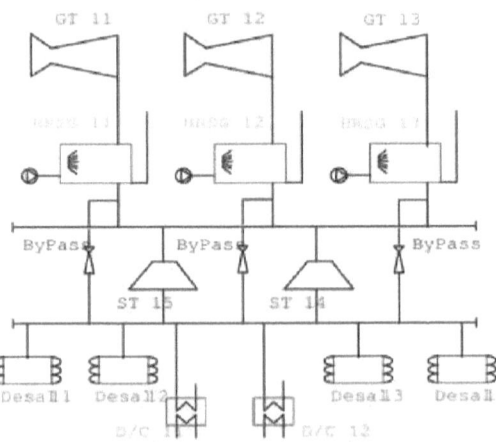

ENERGY AND THE ENVIRONMENT

The conversion of energy from one form to another affects the environment and the air we breathe in many ways.

Fossil fuels such as coal, oil, and natural gas have been powering the industrial development and the amenities of modern life that we enjoy since the 1700s, but this has not been without any undesirable side effects. Such energy conversion affects:

- The soil we farm
- The water we drink
- The air we breathe

ENERGY AND THE ENVIRONMENT

The pollutants released by vehicles and thermal power plants are usually grouped as Hydrocarbons (HC), Nitrogen Oxides (NOx), sulfur dioxide (SO$_2$), sulfur trioxide (SO$_3$), and Carbon Monoxide (CO).

Figure 21

Energy and the Environment

The increase of environmental pollution is happening at alarming rates and the rising awareness of its dangers made it necessary to control it by legislation and international treaties.

POLLUTION EFFECTS

Ozone and Smog:

Smog is the familiar dark yellow or brown haze that builds up in a large stagnant air mass and hangs over populated areas on calm hot summer days.

Smog is made up mostly of ground-level ozone (O_3), but it also contains numerous other chemicals, including carbon monoxide (CO), particulate matter such as soot and dust, volatile organic compounds (VOCs) such as benzene, butane, and other hydrocarbons.

POLLUTION EFFECTS (CONT.)

Ozone and Smog:

The harmful ground-level ozone should not be confused with the useful ozone layer high in the stratosphere that protects the earth from the sun's harmful ultraviolet rays. Ozone at ground level is a pollutant with several adverse health effects.

Ground-level ozone, which is the primary component of smog, forms when HC and NO_x react in the presence of sunlight in hot days.

ENERGY AND THE ENVIRONMENT

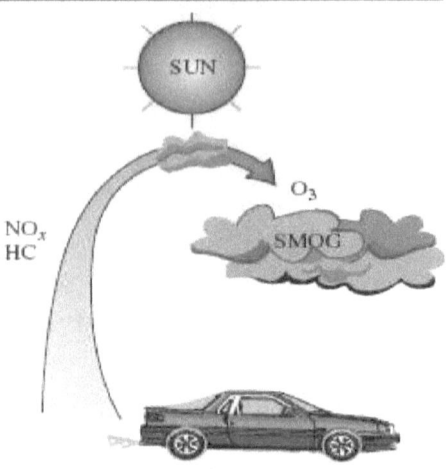

Figure 21 – Formation of smog

ACID RAIN

- Fossil fuels are mixtures of various chemicals, including small amounts of sulfur. The sulfur in the fuel reacts with oxygen to form sulfur dioxide (SO_2), which is an air pollutant.

- The main source of SO_2 is the electric power plants that burn high-sulfur coal. Cars also contribute to SO_2 emissions since gasoline and diesel fuel also contains small amounts of sulfur.

- The sulfur oxides and nitric oxides react with water vapor and other chemicals high in the atmosphere in the presence of sunlight to form sulfuric and nitric acids.

ACID RAIN

Figure 23 – Formation of acid rain

THE GREEN HOUSE EFFECT

The Greenhouse Effect - Global Warming and Climate Change:

We always notice that when we leave our cars under direct sunlight on a sunny day, the interior of the car gets much warmer than the air outside. This is because glass at thicknesses encountered in practice transmits over 90 percent of radiation in the visible range and is practically opaque (nontransparent) to radiation in the longer wavelength infrared regions. Therefore, glass allows the solar radiation to enter freely but blocks the infrared radiation emitted by the interior surfaces. This causes a rise in the interior temperature as a result of the thermal energy buildup in the car. This heating effect is known as the greenhouse effect, since it is utilized primarily in greenhouses.

THE GREEN HOUSE EFFECT

The Greenhouse Effect - Global Warming and Climate Change:

The greenhouse effect is also experienced on a larger scale on earth. The surface of the earth, which warms up during the day as a result of the absorption of solar energy, cools down at night by radiating part of its energy into deep space as infrared radiation. Carbon dioxide (CO_2), water vapor, and trace amounts of some other gases such as methane and nitrogen oxides act like a blanket and keep the earth warm at night by blocking the heat radiated from the earth. Therefore, they are called "greenhouse gases," with CO_2 being the primary

www.ingramcontent.com/pod-product-compliance
Lightning Source LLC
Chambersburg PA
CBHW031616210526
45464CB00004B/1609